Hand Book of
AGRICULTURE

Hand Book of AGRICULTURE

Bharat Singh

ANMOL PUBLICATIONS PVT. LTD.
NEW DELHI - 110 002 (INDIA)

ANMOL PUBLICATIONS PVT. LTD.
4374/4B, Ansari Road, Daryaganj
New Delhi - 110 002
Ph.: 23261597, 23278000
Visit us at: www.anmolpublications.com

Hand Book of Agriculture

First Published, 2006

ISBN 81-261-2559-4

PRINTED IN INDIA

Published by J.L. Kumar for Anmol Publications Pvt. Ltd., New Delhi - 110 002 and Printed at Arora Offset Press, Delhi.

Contents

Preface *ix*

1. **Significance of Land** **1**
Pattern of Usage of Land • Irrigating Areas • Cultivating Areas

2. **Role of Water** **25**
Various Classes • Assessing Water Resources • Expectations in Future

3. **Formation of Soil** **47**
Physical Characteristics • Chemical Characteristics • Organic Elements • Various Soils and Micro-organism • Soil and Mapping Survey • Defining Soil Survey • Drainage of Land • Drainage not on Land • Soil Categorization • Soil Management • Testing Process of Soil • Sample Information Data

4. **Pattern of Cropping** 109
Kharif Cropping Patterns • Cropping Patterns in Kharif Season • Pattern of Cropping in Rabi Period • Process of Crop Planting

5. **Health Problems** 127
Fungicides in Different Types • Lime-sulphur Dilution Table • Fungicides of Other Nature

6. **Role of Climate** 135
Weather Conditions • Differences in Precipitation

7. **Use of Fertilizers** 157
Defining Soil • Soil's Basic Concepts • Categorization of Soil • Marbut's Soil Classification • Soils in India • Soil Formation • Factors of Soil Formation • Soil Profile • Soil Formation in the Process • Components of Soil • Physical Properties of Soil • Dividing Soil Density • Soil Heat : Various Sources • Soil Water Concept • Clay Mineral's Concept • Major Silicate Clay Minerals • Categorization of Silicate Clays • Acidity of Soil • Reclamation of Acidic Soils • Saline and Alkaline Nature of Soil • Important Points of Saline Soil • Reclamation of Saline Soil • Major Elements of Alkaline Soil • Idea of Manures and Fertilizers • Various Fertilizers and their Role • Categorising Nitrogenous Fertilizers • Fertilizers of Phosphatic Nature • Potassium based Fertilizers • Soil Fertility and Productivity • Different Nutrients

8. **Plant Structure** 243

Plant Physiology • Process of Imbibition • Process of Diffusion • Process of Osmosis • Process of Plasmolysis • Coherent Nature • Process of Photosynthesis

9. **Role of Machines** 249

Agriculture and Mechanisation • Ploughs, a Primary Implement • Technique of Bladed Harrows • Use of Soil Implements • Seed-sowing Devices • Harvesting Implements • Thrashing Implements • Farm System of Other Nature • System of Drainage

Bibliography 281

Index 285

Preface

Despite India's intensive desire and all attempts to rise as an industrial nation, the relevance of agricultural sector cannot be underplayed at all. Still one cannot imagine Indian economy sans agriculture. A major share of this sector relies on the weather conditions, so an unfavorable or inclement climate can mar the prospects of the farmers. Only those farmers, who own large-scale farming, can afford the modern facilities of irrigation and mechanization for farming. Very often the farmers find it difficult to preserve their crops and get the desired price for their produce, as most of the small-scale farmers remain illiterate and a disintegrated lot. So the governments at state and central levels need take initiatives for reviving this sector at a revolutionary scale.

This work gives a clear-cut perception of Indian agriculture. All the relevant areas of Indian agriculture have been discussed analytically. Main thrust of the book is on the aspects such as; significance of land, rôle of water, soil formation, cropping patterns, health problems, climatic role, use of fertilizers, plant structure and role of machines, to name a few. In short, it is a complete guide for the farmers, as well as for students of Agriculture Science.

We have been considerably selective about relevant parts of the agriculture sector so this book will certainly make an interesting reading. The reference purpose apart, the general readers may also find it interesting.

—Editor

1

Significance of Land

The pattern of land-use of a country at any particular time is determined by the physical, economic and institutional framework taken together. In other words, the existing land-use pattern in different regions in India has been evolved as the result of the action and interaction of various factors, such as the physical characteristics of land, the institutional framework, the structure of other resources (capital, labour, etc.) available and the location of the region in relation to other aspects of economic development, e.g. those relating to transport as well as to industry and trade. The present pattern can, therefore, be considered to be in some sort of static harmony and adjustment with the other main characteristics of the economy of the region. In the dynamic context, keeping in view the natural endowments and the recent advances in technology, the overall interests of a country may dictate a certain modification of or a change in the existing land-use pattern of a region. A close study of the present land-use patterns and the trends during recent years will help to suggest the scope for planned shifts in the patterns.

Pattern of Usage of Land

Out of the total geographical area of 328 million hectares, the land-use statistics are available for roughly 306 million hectares, constituting 93 per cent of the total. During 1970-71 the latest year for which the land-use data are available, the arable land (the net area sown plus the current and other fallow lands) was estimated at 161.3 million hectares or 52.7 per cent of the total reporting area. Around 65.9 million hectares or 21.6 per cent of the total area was under forests. Land put to non-agricultural uses was estimated at 16.1 million hectares (5.2 per cent of the total) and the barren and unculturable land at 30.2 million hectares or 9.9 per cent of the reporting area. Permanent pastures and other grazing land were estimated at 13 million hectares (4.2 per cent), land under miscellaneous tree crops and groves, not included in the net area sown, at 4.3 million hectares (1.4 per cent) and the culturable wasteland at another 15.2 million hectares or 5 per cent. These figures add up to 306 million hectares of the reporting area.

The area, for which data on the land-use classification are available, is known as the 'reporting area'. In areas where the land-use classification figures are based on land records, the reporting area is the area according to village papers or records maintained by the village revenue agency and the data are based on a complete enumeration of all the areas. In some cases, village papers are not maintained; but the estimates of the area under different classes of land are based on the sample survey or other methods to complete the coverage. The reporting area is the aggregate of the areas based on these two methods. The areas for which no statistics are available are called 'non-reporting area'. The whole of the reporting area is neither completely surveyed cadastrally nor completely covered by complete enumeration of sample surveys. There are still pockets of areas in a few states for which only *'adhoc* estimates' are prepared. Of the total geographical area, only 80.7 per cent is cadastrally surveyed. Of the cadastrally surveyed areas, 91.4 per cent has a permanent reporting agency, whereas 8.6 per cent has no reporting agency.

An idea of the latest position, regarding the 'reporting area' adopting different methods of reporting in different states, can be had from Table.

Area under different systems of reporting, land-utilization statistics—State-wise—1970-71

S. No.	*State/Union Territory*	*Geographical area*	*Reporting area according to the methods of reporting*			
			Complete enumeration	*Sample surveys*	*Adhoc estimates*	*Non-area*
			(Million hectares)			
1.	Andhra Pradesh	27.7	26.4	-	1.0	0.3
2.	Assam	7.9	7.0	-	0.8	0.1
3.	Bihar	17.4	17.3	-	-	0 1
4.	Gujarat	19.6	18.6	-	-	1.0
5.	Haryana	4.4	4.4	-	-	-
6.	Himachal Pradesh	5.6	4.9	-	0.2	0.5
7.	Jammu and Kashmir	22.2*	4.5	-	-	17.7
8.	Karnataka	19.2	18.9	-	-	0.3
9.	Kerala	3.9	-	3.9	-	-
10.	Madhya Pradesh	44.3	44.2	-	-	0.1
11.	Maharashtra	30.8	28.9	-	1.9	-
12.	Manipur	2.2	0.1	-	2.1	-
13.	Meghalaya	2.3	-	-	2.3	-
14.	Nagaland	1.7	-	-	1.4	0.3

Contd...

S. No.	*State/Union Territory*	*Geographical area*	*Reporting area according to the methods of reporting* Complete enumeration	Sample surveys	Adhoc estimates	Non-area
15.	Orissa	15.6	-	15.5	-	0.1
16.	Punjab	5.0	5.0	-	-	-
17.	Rajasthan	34.2	30.4	-	3.7	0.1
18.	Tamil Nadu	13.0	12.8	-	0.2	-
19.	Tripura	1.0	-	-	1.0	-
20.	Uttar Pradesh	29.4	25.6	-	4.2	-
21.	West Bengal	8.8	-	8.9	-	-
22.	Andaman and Nicobar Islands	0.8	0.7	-	0.1	-
23.	Arunachal Pradesh	8.4	-	-	6.4	2.0
24.	Delhi	0.1	0.1	-	-	-
25.	Mizoram	2.1	-	-	2.1	-
26.	Others	0.5	-	-	0.5	-
	All India	328.1	249.8	28.3	27.9	22.1

— = Nil or negligible.

**Includes Dadra, Nagar Haveli, Goa, Daman and Diu Locative, Amindivi and Minicoy Islands and Pondicherry.*

***In some states, the reporting area, according to village papers is slightly more than the geographical area according to the Surveyor-General. As such, the sum of the non-reporting area* in different states does not tally with the all-India non-reporting area.*

****Includes the area under the unlawful occupation of China and Pakistan.*

The non-reporting area is steadily declining from year to year as a result of the special efforts made by the Central government and the state governments, and in 1970-71, it stood at about 7 per cent of the geographical area. This non-reporting area is broadly of two types. First, there are the hill tracts of Jammu and Kashmir and Arunachal Pradesh where there is very little cultivation and where, among other things, owing to the peculiar nature of the terrain, the collection of the annual agricultural statistics is not only very difficult but also costly. Secondly, there are small tracts in some states where owing to the absence of the survey agency or the village revenue agency or both, no regular statistics are collected.

With the extension of cadastral survey to the non-reporting areas and the setting up of the reporting agency in areas where it does not exist, the coverage of land-utilisation statistics is expected to go up. During 1970-71, out of the total non-reporting area of about 22.1 million hectares, about 17.7 million hectares lies in Jammu and Kashmir alone, and the remaining 4.4 million hectares lies in pockets in the states of Andhra Pradesh, Assam, Bihar, Gujarat, Himachal Pradesh, Karnataka, Orissa, Nagaland, Rajasthan and the Union Territory of Arunachal Pradesh. The non-reporting area in Jammu and Kashmir includes the area under the illegal occupation of Pakistan and China, for which agricultural statistics returns are not available.

In the temporarily settled states, there exists a village revenue agency which maintains agricultural statistics as a part of land records. These are collected on the basis of a complete field-to-field enumeration by the *patwari* during the periodical *girdawari* or harvest inspection, and are fairly reliable. On the other hand, in the permanently settled states of West Bengal, Orissa and Kerala, where no such revenue agency exists, agricultural statistics are based on sample surveys.

In the case of some areas where neither complete enumeration nor sample surveys are in vogue, the data are based on the personal knowledge of the local officers, or are based on some old surveys which are not updated every year. The statistics in respect of these areas are, therefore, in the nature of guess-estimates.

Division of Land Utilization : Till 1949-50, the land area in India was classified into five categories known as the five-fold land-utilization classification. These categories were :

(i) forests,

(ii) area not available for cultivation,

(iii) other uncultivated land, excluding the current fallows,

(iv) fallow lands, and

(v) the net area sown. This five-fold classification was, however, a very broad outline of land-use in the country and was not found adequate enough to meet the needs of agricultural planning in the country. The states were also finding it difficult to present comparable data according to this classification owing to the lack of uniformity in the definitions and scope of classification covered by these five broad categories.

To remove the non-comparability and to break up the broad categories into smaller constituents for better comprehension, the Technical Committee on Co-ordination of Agricultural Statistics, set up in 1948 by the Ministry of Food and Agriculture, recommended a ninefold land-use classification replacing the old five-fold classification, and also recommended standard concepts and definitions for all the states to follow.

The statement below gives the nine-fold classification and its relationship with the old five-fold classification.

Classification adopted for land-utilization statistics

S. No.	Old classification	S. No.	New classification
1.	Forests	1.	Forests
2.	Area not available for cultivation	2.	Land put to non-agricultural uses
3.	Other cultivated land, excluding current fallows	3.	Barren and unculturable land
4.	Fallow lands	4.	Permanent pastures and other grazing lands
5.	Net area sown	5.	Miscellaneous tree crops and groves, not included in the net area sown
		6.	Culturable waste
		7.	Fallow land, other than current fallows
		8.	Current fallows
		9.	Net area sown

The total of these classes under both classifications adds up to the reporting area. The revised classification has been accepted in principle by all the states and has been adopted since 1950-51, except by West Bengal, in respect of which the data are still presented on the basis of the old classification.

The above land-use classification is primarily based on whether a particular area is cultivated, grazed or forested. Its main purpose is to show the distribution in detail of the existing land according to its actual use and not how a particular piece of land can be potentially utilized. Thus the area under culturable wasteland does not represent the area which is really culturable, as it may not be possible to bring under cultivation large part of the area, except at a huge cost. Thus the potential land-use classification is beset with several difficulties, as this classification would depend upon the suitability of different areas for different uses, taking into account their natural endowments, the availability of capital and other resources for the development of the land for the desired use and the likely economic returns. If the potential use of the land has to be taken into account, a large amount of data relating to the inherent characteristics of each soil type and the economics of putting it to a particular use would have to be specially collected through soil surveys, land-use surveys and wasteland utilization surveys.

On the basis of the above nine-fold classification, it is possible to build up the old five-fold classification as well as to arrive at the area according to concepts like 'arable land' or 'potential land' available for crop husbandry. The 'arable land' would comprise the 'net area sown' plus the 'current fallows' and 'other fallow land'. Similarly, the 'potential land' available for cultivation would include, besides the 'arable land', the land under 'culturable waste', 'permanent pastures and grazing land', and 'miscellaneous tree crops and groves, not included in the net area sown'. However, such potential land cannot be ascertained without surveys of the soil types and land-use and the economics of bringing them under cultivation. The utility of the above concepts of 'arable' and 'potential' land can be enhanced if some indicators of arable and potential land are computed:

(i) Percentage potential land exploited $= \frac{\text{Net area sown}}{\text{potential land}} \times 100$

(ii) Percentage arable land exploited $= \frac{\text{Net area sown}}{\text{arable land}} \times 100$

With the adoption of the nine-fold classification since 1950-51 an element of non-comparability has been introduced in the data before and after that year. For instance, in the old land-utilisation classification, the term 'current fallows' included the land lying fallow even up to a period of 10 years in the former Bombay State and for 2 years in the former Punjab State, whereas in the revised nine-fold classification, the current fallows have been limited to the lands lying fallow for one year only, and the term 'other fallow land' includes land lying fallow for more than one year, but less than five years. Thus the area under 'current fallow' in the old five-fold classification need not necessarily add up to two sub-classes in the new classification, i.e. 'current fallows' and 'other fallow land'. Some of the land lying fallow beyond five years may have been included in the nine-fold classification as 'culturable waste'.

To standardize the concepts and definitions, the Technical Committee on Co-ordination of Agricultural Statistics laid down standard definitions of the various categories of land use.

Based on the nine-fold classification, the all-India data on land utilization for 1970-71 are given below :

Land utilisation in India—1970-71

Classification	*(Area '000 ha)*	*Percentage of the reporting area*
1	2	3
1. Total geographical area	328,048	-
2. Reporting area : 305,985	100.0	
(i) Forests	65,928	21.6
(ii) Not available for cultivation	46,215	15.1
(a) Non-agricultural uses	16,049	5.2
(b) Barren and unculturable	30,166	9.9
(iii) Other uncultivated land (excluding fallow land)	32,500	10.6
(a) Permanent pastures and other grazing land	12,996	4.2
(b) Miscellaneous tree crops and groves	4,339	1.4
(c) Culturable wasteland	15,165	5.0

Contd...

Classification	*(Area '000 ha)*	*Percentage of the reporting area*
1	2	3
(iv) Fallow land 20,181	6.6	
(a) Fallow land other than current fallows	9,072	3.0
(b) Current fallows	11,109	3.6
(v) Net area sown	141,161	46.1
(vi) Gross cropped area	167,412	-
(vii) Area sown more than once	26,251	-
(viii) Net irrigated area	31,292	-
(ix) Gross irrigated area	38,552	-

In many countries, there is considerable scope for bringing new areas under cultivation, but in India for the extension of cultivation to new land is very limited Already about 49.7 per cent of the total reporting area is cultivated. Though culturable land which is not cultivated at present (culturable wasteland and other follow lands and permanent pastures and grazing lands and miscellaneous tree crops), is estimated at about 42 million hectares of 13.6 per cent of the total, most of this area is occupied by Culturable wastelands. Before 1946 there were no hectares of weed-infested land and jungle which could be cleared and made available for cultivation for growing additional foodgrains. Therefore the Government of India through the Central Tractor Organization (CTO) started operations in 1947-48 to reclaim about one million hectares of wasteland within a period of five years. Till October Uttar Pradesh, Madhya Pradesh, Assam, Madras and Punjab states were reclaimed. Jungle clearance was under taken in the states of Uttar Pradesh, Madhya Pradesh and Assam.

The C.T.O. was wound up, as the state governments wanted to reclaim land themselves, instead of hiring the services of the C.T.O., other problems of follow-up cultivation of reclaimed lands also came up.

In June 1959, the Government of India constituted a Committee to make a survey of land classified as 'other uncultivated land, excluding fallow lands' and 'fallow lands other than current follows' and to locate areas where large blocks of lands were available for reclamation and resettlement. By 1963, the Committee completed its survey in 13 states of Punjab, West Bengal, Bihar, Mysore, Uttar Pradesh, Madhya Pradesh, Kerala, Madras, Andhra Pradesh, Jammu and Kashmir, Orissa and Gujarat.

The Committee estimated that in these states, about 0.8 million hectares of wasteland was available for cultivation in blocks of 100 hectares (or 250 acres) or more. The total wasteland in such large blocks in the whole country was estimated at 1.2 million hectares. Typical wastelands in each state were visited by the Committee and were classified according to certain broad criteria, namely those affected by water-logging, salinity and alkalinity, soil erosion, and infestation with thick growth of bushes.

After studying the type of soil, rainfall, natural vegetation, etc. suitable reclamation measures were suggested for bringing the areas under cultivation. In addition, wherever necessary, resettlement schemes needed for keeping the areas under cultivation were also proposed. Suitable cropping patterns for various areas and the economics of reclamation in respect of these areas were discussed.

Finally, priorities were indicated among various categories of reclaimable lands. In regard to the data on culturable wastelands, the Committee observed that land classified as culturable waste at the time of settlement sometimes continued to be shown as such in the revenue records long after they had come under cultivation.

In the view of the Committee, the mere collection of statistics under the head 'culturable waste' served little purpose and detailed information about the types of wasteland in each state, the ownership of such land, its availability in sizeable blocks and the cost of reclamation measures should be collected. The Committee, therefore, recommended that rapid reconnaissance surveys should be conducted for collecting such information. The Committee further recommended the survey and categorization of wastelands in blocks measuring less than 100 hectares. For this purpose, several schemes were taken up by different state governments during the Third Plan.

Taking into account the total land resources including hills, mountains, lakes, rivers and lands of all descriptions, the availability of land per head in India comes to only 0.58 hectare as compared with the availability of 59 hectares per head in Australia, 45.07 hectares in Canada, 9.06 hectares in the USSR, 4.48 hectares in the USA, 2.33 hectares in Burma, 1.21 hectares in Pakistan, 0.43 hectare in the UK and 0.35 hectare in Japan. It is clear that the availability of land per head in India is among the lowest in the world. This is a natural corollary to the population explosion in old countries, such as India.

The percentage of arable land to the geographical area in India works out at 52.7 as compared with 31.8 in France, 23.9 in Pakistan, 29.6 in the UK, 27.4 in Burma, 20.4 in the USA, 13.0 in Japan, 10.2 in the USSR and 5.8 in Australia.

The proportion of area under agricultural uses (arable land, permanent pastures and grazing land and land under miscellaneous tree crops and groves) to the total geographical area in India is estimated at, 58.4 per cent as compared with

77.2 per cent in the UK, 46.6 per cent in the USA, 30.1 per cent in Pakistan, 28.5 per cent in Burma, 27.1 per cent in the USSR and 17.2 per cent in Japan.

The total area under all crops or the gross cropped area, counting the areas sown more than once during a year as many times as they are sown, in India during 1970-71, was 167.41 million hectares. The difference between the gross cropped area and the not area sown represents the area sown more than once and this was estimated at 26.25 million hectares.

Thus the intensity of cropping representing the percentage of the gross cropped area to the net area sown works out at 118.6.

Forest Cover Area : According to the land-utilization statistics, the area under forests during 1970-71 is estimated at 65 9 million hectares. However, according to the returns of forest statistics furnished by the state forest departments, the forest area is put at 75.8 million hectares.

The difference in the two sets of figures is due mainly to the variations in the definitions of the term 'area under forests', in geographical coverage and in the reference period. The question of reconciliation of the two sets of figures has been engaging the attention of the Central government and the state governments.

A uniform definition has now been adopted for reporting the data by both the sources and it is hoped that the gap between the two sets of figures would be narrowed down considerably, if not altogether eliminated. In India, forests account for about 22 per cent of the total land surface. On the other hand, in advanced countries, the area under forests is often about a third of the total land area. Table gives the estimates of geographical, arable and forest areas per head in different countries of the world.

Availability of geographical, arable and forest areas in various countries per head—1972

S. No.	Country	Hectares per head			Percentage arable land to the geographical area	Percentage of the forest area to geographical area	Percentage of the agricultural area to the geographical area
		Geographical area	Arable land	Forest area			
1		2	3	4	5	6	7
1.	Australia	59.00	3.45	2.68	5.8	4.5	64.4
2.	Canada	45.07	1.98	20.00	4.4	44.4	6.9
3.	U.S.S.R.	9.06	0.92	3.68	10.2	40.6	27.1
4.	Brazil	8.65	0.30	5.26	3.5	60.8	16.1
5.	U.S.A.	4.48	0.91	1.46	20.4	32.7	46.6
6.	Egypt	2.79	0.08	Negligible	2.7	Negligible	2.8
7.	Burma	2.33	0.64	1.34	27.4	57.5	28.5
8.	China (Main land)	1.20	0.13	0.12	11.6	10.0	30.0
9.	France	1.06	0.34	0.27	31.8	25.6	60.4
10.	Poland	0.94	0.45	0.26	47.9	27.4	62.4
11.	Pakistan*	1.21	0.29	0.03	23.9	2.3	30.1
12.	India**	0.56	0.29	0.12	52.7	21.5	58.4
13.	Italy	0.55	0.17	0.11	31.6	20.5	58.6
14.	U. K.	0.43	0.13	0.03	29.6	8.0	77.2
15.	F.D.R. Germany	0.42	0.13	0.12	30.5	27.0	54.6
16.	Japan	0.35	0.05	0.24	13.0	69.0	17.2
17.	World average	3.56	0.39	1.07	10.9	30.2	33.2

**Based on population figure as on March 1, 1971 and the land-utilization data for 1970-71.

Source : F.A.O. Production Yearbook, 1972.

It has been observed that considerable portions of Indian forests are such only in name and are subject to various forms of maltreatment. As a result, a hectare of forest land in India yields less utilizable timber than a hectare of forest in Europe. Having regard to these considerations, specially the lower productive capacity of the natural tropical forests in India, the National Forest Policy Resolution of 1952 proposed that the area under forests be raised steadily to 33 per cent of the total area, the proportion to be aimed at being 60 per cent in the hill regions and 20 per cent in the plains. It would be worthwhile to get an idea of the trends in the land-use patterns over the last two decades. A comparative picture of the land-use patterns during 1950-51 and 1970-71 is presented in Table.

Owing to extension in the coverage of the land-utilization statistics and changes in the concepts and definitions adopted for the classification, the two sets of figures are not strictly comparable. However, it can be broadly inferred that there has been an increase in the area under forests, land under non-agricultural use, permanent pastures and grazing lands and the net area sown. This led to a decline in the barren and unculturable land, land under miscellaneous tree crops and groves, culturable wastelands and fallow lands.

All India land-use classification during 1950-51 and 1970-71.

	Item	*Percentage of the reporting area in*	
		1950-51	*1970-71*
1.	Land under forests	14.2*	21.6
2.	Land not available for cultivation	16.7	15.1
2.1	Land under non-agricultural uses	3.3	5.2
2.2	Barren and unculturable land	13.4	9.9
3.	Other uncultivated land, excluding fallow land	17.4	10.6
3.1.	Permanent pastures and other grazing land	2.3	4.2

Contd...

Item	*Percentage of the reporting area in*	
	1950-51	*1970-71*
3.2. Land under miscellaneous tree crops and groves, not included in the net area sown	7.0	1.4
3.3. Culturable waste	8.1	5.0
4. Fallow land	9.9	6.6
4.1 Fallow land other than current fallow	6.1	3.0
4.2. Current fallow	3.8	3.6
5. Net area sown	41.8	46.1
6. Reporting area	1 00.00	100.00

*A sizeable portion of the non-reporting area consisted of the forest area. As such, the percentage of area under forests to the total reporting area was somewhat low cultural raw materials. The demand was partly met through the extension of the area under cultivation and partly through intensive cultivation. During the First and Second Plans, the increase in overall agricultural production resulted largely from the extension of the area, whereas during the Third Plan and thereafter, the increase in productivity contributed more to increased production. An idea about these trends can be had from Table which shows the compound annual growth rates of production, area and productivity of all crops over the last two decades. During the last two decades, there have been some shifts in the areas from coarse foodgrains to superior cereals, such as wheat, or from foodgrains to certain commercial crops. The percentage area under pulses and foodgrains, in general, to the total cropped area showed marginal declines. On the other hand, the area under commercial crops, such as oilseeds, fruits, vegetables, cotton and jute showed some increase.

Compound rates of growth in area, production and productivity for all crops

Plan periods	*Area*	*Production (Percentage*	*Productivity per annum)*
First Plan @ (1951-56)	2.6	-4.1	1.4
Second Plan @ (1956-61)	1.3	3.1	1.8
1961-73*	0.4	2.0	1.3

Source : @The Fourth Five-Year Plan (1969-74)

*Indian Agriculture in Brief (Thirteenth Edition)

These shifts in the cropping patterns are presented in Table.

The cropping patterns in India

Crops	*Percentage of the gross cropped area in*				
	1950-51	*1955-56*	*1960-61*	*1965-66*	*1970-71*
Rice	23.6	21.5	22.3	22.8	22.3
Wheat	7.6	8.6	8.5	8.1	10.8
Jowar	11.8	11.8	12.1	11.3	10.2
Cereals (total)	61.1	59.5	60.2	59.3	60.7
Pulses (total)	15.6	16.0	15.5	14.7	13.7
Foodgrains (total)	76.7	75.5	75.7	74.0	74.4
Sugarcane	1.5	1.3	1.6	1.8	1.6
Fruits (total)	0.5	0.6	0.7	0.9	0.9
Vegetables (total)	1.1	0.9	1.0	1.2	1.3
Oilseeds (total)	8.3	8.1	8.4	9.1	8.8
Cotton	4.3	5.7	5.0	5.1	4.7
Jute	0.4	0.5	0.4	0.5	0.4
Tea	0.2	0.2	0.3	0.2	0.2
Coffee	0.07	0.08	0.2	0.1	0.1

Irrigating Areas

Intensive farming and multiple-cropping need irrigation. In India, the statistics of irrigation are collected along with the land-utilization statistics which are published annually. Information regarding the total area irrigated from different sources of irrigation and the area under different crops irrigated during autumn, winter and summer is available.

The total area irrigated under all crops represents the gross irrigated area which comprises the net irrigated area plus the area on which two crops are grown and irrigated during the same agricultural year. About four-fifths of the country's cropped area depends exclusively on rainfall, most of which is concentrated in a few months of the year. Even where the overall annual precipitation is high, the available moisture during winter and summer is not adequate to support multiple-cropping.

The rainfall in large parts of the country is so low and uncertain in its distribution that it does not permit intensive cultivation even during the main crop season. The expansion of irrigation facilities in order to ensure timely and adequate water supply has, ever since the inception of planning, been an extremely important means of bringing about agricultural development in the country.

The net area irrigated in India during 1970-71 was 31.3 million hectares which was about 22.2 per cent of the net area of 141.2 million hectares sown during that year. The area irrigated more than once was 7.3 million hectares of 23.2 per cent of the net area irrigated. The gross irrigated area worked out at 38.6 million hectares. Table presents the trends in the net area an the gross area irrigated and the area irrigated more than once.

Table : The all-India estimates of the net area sown, gross cropped area, area grown more than once and the irrigated area over 1950-51 to 1970-71.

	Item	*1950-51*	*1955-56*	*1960-61*	*1965-66*	*1970-71*
	(Area in thousand/hectares)					
1.	Net area sown	188,746	129,156	133,199	136,242	141,161
2.	Gross cropped area	131,893	147,311	152,772	155,364	167,412
3.	Area sown more than once	13,147	18,155	19,573	19,122	26,251
4.	% of double-cropped area to the net area sown	11.1	14.1	14.7	14.0	18.6
5.	Net irrigated area	20,853	22,758	24,661	26,343	31,292
6.	Gross irrigated area	22,563	25,642	27,980	30,879	38,552
7.	Area irrigated in more than one season	1,710	2,884	3,319	4,536	7,260
8.	% of area irrigated in more than one season to the net irrigated area	8.2	12.7	13.5	17.2 23.2	
9.	Double-cropped area with no irrigated crop or just one irrigated crop (3-7)	11,437	15,271	16,254	14,586	18,991
10.	Item 9 as percentage of the total double-cropped area	87.0	84.1	83.0	76.0	72.0

It is evident from Table that the area sown more than once was below 15 per cent up to 1965-66. It is only during the last few years that there has been some increase in such areas because of the adoption of short-duration varieties of cereals and as a result, the area sown more than once has gone up to 18.6 per cent in 1970-71.

Of the gross irrigated area, which accounts for 38.6 million hectares, about 79.3 per cent or 30.6 million hectares is utilized for the production of foodgrain crop, as will be seen from Table.

Table : Irrigated area under foodgrain and non-foodgrain crops in India—1950-51 to 1970-71.

Crops	*1950-51*	*1960-61*	*1965-66*	*1970-71*
	(Area in million hectares)			
Foodgrains	18.3	22.1	24.0	30.6
	(81.3)	(78.9)	(77.8)	(79.8)
Non-foodgrains	4.3	5.9	6.9	8.0
	(18.7)	(21.1)	(22.2)	(20.7)
All crops	22.6	28.0	30.9	38.6
	(100.0)	(100.0)	(100.0)	(100.0)

Note: Figures in parenthesis show the percentages of the gross irrigated area.

The above all-India comparisons conceal a large unevenness in the distribution of the land-utilization patterns prevailing in the constituent states and union territories. The states that report agricultural statistics for their entire geographical area are Haryana, Kerala, Madhya Pradesh, Maharashtra, Manipur, Meghalaya, Orissa, Punjab, Tamil Nadu, Tripura, Uttar Pradesh, West Bengal and the Union Territories of Andaman and Nicobar Islands, Delhi and Mizoram. As already referred to the states of Jammu and Kashmir, Gujarat, Himachal Pradesh, Andhra Pradesh, Karnataka, Nagaland and the Union Territory of Arunachal Pradesh have some non-reporting areas.

Land put to non-agricultural uses. During the recent years, there has been an increase in the area put to non-agricultural uses, as expected, because as a result of increase in the development activities, more and more land is being used for industrial sites, housing, transport systems, recreational purposes, irrigation systems, etc. States where the proportion of lands under non-agricultural uses is higher than the all-India average are Haryana, Jammu and Kashmir, Kerala, Orissa, Uttar Pradesh, Andhra Pradesh. Punjab, Tamil Nadu, Bihar, Assam and the Union Territories of Goa, Daman and Diu, Delhi and Pondicherry. The states which jointly account for more than two-thirds of the land under non-agricultural uses are Andhra Pradesh, Madhya Pradesh, Uttar Pradesh, Bihar, Tamil Nadu, Rajasthan, Orissa and Karnataka.

The other types of areas, which are covered under barren and unculturable lands, are generally unsuitable for agricultural use either because of the bad soil and topography or because of their inaccessibility. Instances are the desert areas in Rajasthan, the saline lands in parts of the Rann of Kutch in Gujarat, the weed-infested and ravine lands in Madhya Pradesh and *usar* or alkaline lands in Uttar Pradesh. The proportions of barren and uncultivated lands to the reporting areas are higher in the states of Rajasthan, West Bengal, Assam, Gujarat, Manipur, Nagaland, Meghalaya and the Union Territories of Arunachal Pradesh and Mizoram. The states of Rajasthan, Gujarat, Uttar Pradesh, Madhya. Pradesh, Andhra Pradesh, Meghalaya, Assam and Maharashtra together account for more than 67 per cent of the land under this category in the country.

The other uncultivated land, excluding current fallows, covers areas classified under permanent pastures and grazing lands, the areas under miscellaneous tree crops and groves and culturable wastelands. The states which have considerable proportions of areas under permanent pastures and grazing lands are Himachal Pradesh, Karnataka, Madhya Pradesh, Gujarat, Rajasthan, Tripura and the Union Territory of Dadra and Nagar Haveli. Madhya Pradesh accounts for about 24 per cent of the all-India acreage under pastures and grazing lands. About one-fourth of the country's land under miscellaneous tree crops and groves lies in Uttar Pradesh. The states of West Bengal, Orissa, Andhra Pradesh, Karnataka and Tamil Nadu also have sizeable areas under this category.

The states with substantial proportions of areas under culturable wastelands are Rajasthan, Orissa and the Union Territory of Goa, Daman and Diu. The states of Rajasthan, Madhya Pradesh, Uttar Pradesh, Andhra Pradesh and Orissa together account for more than three-fourths of the nation's land resources under this category. Rajasthan alone accounts for over two-fifths of the nation's culturable wasteland.

Of the country's total land under current fallows, the states of Rajasthan, Andhra Pradesh, Bihar, Maharashtra, Tamil Nadu, Karnataka, Uttar Pradesh, Madhya Pradesh, Orissa and Gujarat account for more than 90 per cent.

Land lying fallow for more than one agricultural year but less than five agricultural years has decreased during the recent years. This decrease is probably due to the extension of cultivation to marginal lands as a result of the increased demand for agricultural produce, development of irrigation, etc. Still their proportions continued to remain substantial in the states of Rajasthan, Bihar, Tamil Nadu, Maharashtra, Andhra Pradesh, Karnataka, West Bengal and the Union Territory of Delhi. These states account for over three-fourths of the country's area under 'other fallow lands'.

Cultivating Areas

The states of Haryana, Punjab, West Bengal, Maharashtra, Uttar Pradesh, Kerala, Karnataka, Gujarat, Bihar, Tamil Nadu and the Union Territories of Laecadive, Minicoy and Amindiye Islands (Lakshadweep), Pondicherry and Delhi cultivate more than 45 per cent of their reporting area. In some of the states like Haryana and Punjab, which lie in the fertile Indo-Gangetic plains of India, the net area sown is as high as 80 per cent of the reporting area ; whereas in the states of West Bengal and Maharashtra and the Union Territory of Pondicherry, the net area sown is between 60 and 70 per cent of the corresponding reporting area. Maharashtra has the highest net area sown in the country. The states of Maharashtra, Madhya Pradesh, Uttar Pradesh, Rajasthan, Andhra Pradesh, Karnataka, Gujarat and Bihar account for more than three-fourths of the country's net area sown.

The gross cropped area is the highest in Uttar Pradesh and is followed by that in Madhya Pradesh, Maharashtra, Rajasthan, Andhra Pradesh, Bihar, Karnataka, Gujarat, Orissa, Tamil Nadu, West Bengal, Punjab, Haryana, Kerala, Assam, Himachal Pradesh, Jammu and Kashmir, Tripura, Meghalaya, Manipur and Nagaland. The percentage of the areas sown more than once of the net area sown is greater than the all-India average of 18.6 per cent in the states of Punjab, Haryana, Uttar Pradesh, West Bengal and Bihar.

Uttar Pradesh has the largest gross irrigated area in the country or 21.6 per cent of the country as a whole. It is followed by Punjab with 11.0 per cent and Andhra Pradesh 10.95 per cent. In the case of foodgrains, Uttar Pradesh again has the highest irrigated area (22.3 per cent) in the country, and is followed by Andhra Pradesh (11.7 per cent) and Punjab (9.7 per cent). In the case of non-foodgrain crops, Uttar Pradesh and Punjab account for 19.2 and 16.1 per cent respectively, of the total irrigated area under these crops.

In India, foodgrains occupy a predominant position in most of the states, accounting for more than two-thirds of the gross cropped area. In fact, the states of Bihar, Madhya Pradesh, Uttar Pradesh, West Bengal, Himachal Pradesh, Jammu and Kashmir, Manipur and Nagaland have more than 80 per cent of the gross cropped area under foodgrains. In the states of Kerala and Gujarat, leas than 50 per cent of the gross cropped area is devoted to foodgrains.

Foodgrains account for more than 70 per cent of the gross irrigated area in all states, except Gujarat, Maharashtra and Punjab. In Gujarat, more than 50 per cent of the total irrigated area is under non-foodgrain crops.

The available agricultural land in the country is already being put to some agricultural use and in each region, a certain pattern of land utilization has already been evolved. Keeping in view the increasing demand for various agricultural, forestry and livestock products for internal consumption and exports, it is very necessary that each piece of land is put to its best use. Further, as the various development programmes, such as soil conservation, the spread of irrigation facilities and the adoption of improved agricultural practices, are introducing an element of dynamism in the land-use and cropping patterns, it would be expedient to adopt a planned approach to land utilization with the following objectives : (i) The development of all arable land for agriculture or its restoration to agriculture where necessary, keeping in view the requisite balance among farming, pastures and forestry; (ii) the achievement of an apportionment of land between different uses and a combination of land with other factors (e.g. intensity of cultivation) which would help to produce the needed quantity and quality of different agricultural or forestry products at the lowest possible costs in terms of the factors of production employed in or available to agriculture; (iii) the improvement in the quality of used land through such measure? as irrigation and conservation of the fertility of land ; and (iv) to facilitate such shifts in land-use as would increase the returns to the factors employed in agriculture and thus lead to an improved pattern of production without causing any serious dislocations in the economy.

For achieving these objectives, proper land-use planning is a prerequisite. Before the land-use planning is done, it is necessary to have a detailed classification of the country into agro-climatic regions and zones, based on the data on rainfall, temperature, humidity, soils, existing cropping patterns, irrigation, density of population, livestock, etc. After such a classification is done, it would be necessary to work out the optimum land-use and cropping patterns for each area, consistent with the overall demands for agricultural commodities to satisfy the domestic and export requirement. This should form the basis of agricultural development programmes under the successive Five-Year Plans.

2

Role of Water

India is rich in water resources, being endowed with a network of great rivers and vast eluvial basins to hold groundwater. Conditions, however, vary widely from region to region. Whereas there are some chronically drought-affected areas, there are others which are frequently subject to damage by floods.

On the whole, under the pressure of rapid population growth, the available resources of water are being developed and depleted at a fast rate and the situation seriously underlines the need for taking up integrated plans for water conservation and utilization for every agro-ecological area to meet the increasing demands of irrigation, water-harvesting, human and livestock consumption, expanding industry, hydro-electric power generation, recreation, navigation and other uses.

Various Classes

Water resources are divisible into two distinct categories: the surface-water resources and the groundwater resources.

Each of these categories is a part of the earth's water circulatory system, called the hydrologic cycle, and is ultimately derived from precipitation, which is rainfall plus snow. They are interdependent and frequently the loss of one is the gain of the other. The brief description of the runoff cycle, which is a part of the hydrologic cycle, will help us to understand the origin and the interdependence of these two categories of water resources. The precipitation that falls upon land and is the ultimate source for both the categories of water recourses is dispersed in several ways. A sizeable portion is intercepted by the vegetal cover or temporarily detained in surface depressions. Most of it is later lost through evaporation. When the available interception or the depression storage are completely exhausted and when the rainfall intensity at the soil surface exceeds the infiltration capacity of the soils, the overland flow begins. Once the overland flow reaches a stream channel, it is called surface runoff, which, together with other components of flow, forms the total runoff.

Part of the water that infiltrates into the surface soil may continue to move laterally at shallow depth as interflow owing to the presence of relatively impervious lenses just below the soil surface and may eventually reach the stream channel when it is called the sub-surface runoff. A part of the sub-surface runoff may enter the stream promptly, whereas the remaining part may take a long time before joining the stream flow.

A second part of the precipitation which infiltrates is lost through evapotranspiration *via* plant roots and thermal gradients just below the soil surface. A third part may remain above the water-table in the zone of unsaturated flow. A fourth remaining part percolates deeply into the groundwater. Part of this groundwater may eventually reach the stream channel and become the base flow of the stream. This portion is termed groundwater runoff or groundwater flow.

Apart from infiltrated rain-water, the seepage from canals, ponds, tanks, lakes, irrigated fields, etc. is also dispersed and accounted for in the same manner.

The total runoff in the stream channel includes the snowmelt, the surface run off, the sub-surface runoff, the groundwater runoff and the channel precipitation, i.e. the precipitation falling directly on the water surface of streams, lakes, etc. It constitutes what is known as the surface-water resources. The portion of the precipitation which, after infiltration, reaches the ground-water-table, together with the contribution made to groundwater from a neighbouring basin, influent rivers, natural lakes, ponds, artificial storage reservoirs, canals, irrigation, and constitutes the groundwater resources. That quantity of water in the groundwater reservoir, which is not annually replenishable, is not taken into account, as it is a sort of dead storage which cannot be used on a continuing basis from year to year.

The above phase of the runoff cycle pinpoints the inflow components for the surface—as well as for the groundwater resources. It has to be appreciated that there is always a balance between the inflow factors making up water resources of a region, whether surface or ground, and the outflow components. The surface-water resource of a given basin in excess of the withdrawal use is accounted for on the outflow side by one or more of the following factors : (i) Stream outflow from the basin; (ii) loss through evaporation; and (iii) the influent recharge to the groundwater.

Similarly, the unutilized groundwater resource of a basin is accounted for by the following outflow factors: (i) Evapotranspiration from the ground-water-table; (ii) outflow to the neighbouring groundwater basin : (iii) the effluent discharge to the streams; and (iv) the addition to the groundwater storage.

The interrelationship between the surface-water and the

groundwater resources is evident from the above analysis. The surface-water resources contribute to the groundwater recharge in various ways : (i) by influent recharge from the streams ; (ii) by seepage from natural lakes, ponds, etc.; (iii) seepage from artificial storage reservoirs, canal systems, etc., and (iv) return flow from irrigation. These factors presently contribute to about 25 per cent of the country's total groundwater resources.

On the other hand, the bulk of the base-flow in the rivers, which represents the sustained fair-weather runoff is contributed by the groundwater resources. This contribution, presently, is roughly assessed at about 25 per cent of the total surface-water resources of the country.

Role of Water Resources and the Factors Affecting **:** The water resources of a region, conceived as a dynamic phase of the hydrologic cycle, are influenced by the following three major groups of factors :

1. *Factors Regarding Climate*

(A) Rainfall : its intensity, duration and distribution.

(B) Snow

(C) Evapotranspiration.

2. *Factors Regarding Physiography*

(A) Basic characteristics.

1. Geometric factors : drainage area, shape slope and stream density.
2. Physical factors : land use, surface infiltration conditions, soil types, etc.

(B) Channel characteristics: carrying capacity and storage capacity.

3. *Factors Regarding Geology*

(A) Lithologic including composition, texture, sequence of rock types and the thickness of rock formations.

(B) Structural, including chief faults and folds that interrupt the uniformity of occurrence of rock types or sequence of rock types also beds, joints, fissures, cracks, etc.

(C) Hydrologic characteristics of the aquifers permeability, porosity, transmissivity, storability, etc.

The physiographic features (including geological factors) not only influence the occurrence and distribution of water resources within a region but these, particularly the orography, play a significant role in influencing rainfall and other climatic factors, such as temperature, humidity and wind. However, within a given geographical location and physiographic framework, it is primarily the rainfall (its intensity, duration and distribution) and the climatic factors affecting evapotranspiration that determine the totality of water resources in the region.

Precipitation and Distribution : Rainfall pattern and distribution in a region is a good index of its water resources.

A considerable portion of the Indian subcontinent belongs to the subtropical zone. However, the region as a whole shares the characteristics of a tropical climate. Literally, the word 'monsoon' means a wind system which undergoes a seasonal 180° reversal of direction. Many regions of the world experience the monsoon. In India, however, two factors make it unique. One is the continuous and high mountain mass in the north which forms an effective barrier to the air movement across them. The second is the peninsular shape of the subcontinent with its land in close proximity to the ocean, thereby providing a rich source of moisture.

The rainfall in the country is primarily orographic, associated with tropical depressions originating in the Bay of Bengal and the Arabian Sea. The moisture-laden summer monsoon, accounting for the bulk of rainfall in the country, originates from the vast expanse of the Indian Ocean and enters the Indian subcontinent from the south-west as south-westerly current. The physiographic features of the Indian Peninsula and the Western Ghats divert the monsoon into two branches namely, the Arabian Sea branch and the Bay of Bengal branch. The Arabian Sea branch strikes the Western Ghat and precipitates heavily along the Western Ghat from Kerala to Gujarat between the last week of May and the first week of June. After surmounting the Ghats the southern part of the current blows across the Peninsula as a westerly or, in places, as a north-westerly wind. The northern portion of the current which crosses the Saurashtra coast blows across Rajasthan as a south-westerly wind and gives rain mostly in the coastal districts near the Aravalli Hills and the Punjab Kumaon Hills, but very little in the plains of Rajasthan.

The Bay of Bengal branch turns north and enters Bangladesh, Assam and West Bengal in May as a southerly current. With the progress of summer, this current moves northwards and by about June it strikes the Himalayas and precipitates heavily. On being deflected by the Himalayas, it turns westwards and enters northern India as an easterly current. This occurs in the whole country, except in the western parts of Rajasthan, coming under the influence of the monsoon by about the first week of July. However, the advance of monsoon is not regular every year and is governed by the movement of depressions and low-pressure waves travelling from the Bay of Bengal across the country. Whereas Assam, the western coastal plains and the Western Ghats receive heavy rainfall almost every year, rainfall in other parts of the country is dependent on the number of depressions in the Bay of Bengal. Therefore, the annual monsoon variation in the central, northern and the north-western parts of the country is much more pronounced than in the regions of abundant rainfall

The monsoon starts withdrawing from Punjab downwards

up to Saurashtra and Gujarat in the first week of September and disappears by the middle of November, except in the extreme south and south-east, where the receding monsoon gives rain as the north east monsoon from the Bay of Bengal.

During winter, the northern part of the country gets some rainfall from western disturbances, but these are irregular and not reliable compared with the south-west monsoon. Also, severe cyclonic storms experienced during the transition months of April to June and October to December cause some precipitation. Some parts of the country receive hot-weather rainfall between March and May mainly owing to large-scale thunderstorms called Norwesters. These rains are of substantial importance in West Bengal and Assam.

The annual average rainfall in India is about 120 cm, i.e. slightly more than the global mean of 99 cm. The more important fact to reckon with, however, is its extreme spatial and temporal variation. The regional variation is so large that whereas the Khasi Hills in the north-east get as average annual rainfall of more than 1,000 cm, the average annual rainfall received in parts of the Rajasthan desert is even less than 15 cm. Moreover, nearly 75 per cent of the total rainfall is received in the four-month period from June to September. Even in this season, the rainfall in most parts of the country is subject to uncertainty of occurrence, marked by prolonged dry spells and aberration in the time of commencement and withdrawal, and also the total amount received. This explains the dependence of agriculture in most parts of the country on artificial irrigation and gives an insight into the nature of the country's water resources.

Evaporation and evapotranspiration. Evaporation is the process by which water vapour escapes from a free water-surface or moist soil. On land, in addition to water lost from the soil, there is also loss of water from plants; the combined loss is termed 'evapotranspiration'. Evapotranspiration is an important parameter in determining the water resources of a region. Loss through evapotranspiration is controlled mainly by two factors, namely, the ability of the atmosphere to supply energy to vaporize water and transport the vapour and enhance the availability of moisture on the earth's surface. The former is determined by such parameters as the short wave and the long wave radiation, humidity, air temperature, wind movement, albedo and temperature of the evaporating surfaces. The availability of moisture at the earth's surface is influenced by precipitation and the artificial application of water. Evapotranspiration is also influenced to some extent by such factors as vegetal cover, type of foliage, stage of plant growth, and soil fertility. Because of all these variable complex factors, the estimation of evapotranspiration becomes very difficult.

A number of theoretical and empirical methods, however, have been used for computing the evapotranspiration. These include methods based on the energy-budget theory, mass-transfer theory, water-budget approach and empirical methods using various equations, the prominent among which are : (1) Lowry-Johnson; (2) Thornthwaite; (3) Blaney-Criddle; (4) Blaney-Morin; (5) Penman; (6) Hargreaves; and (7) Christiansen.

Whereas some of these equations compute the evapotranspiration directly, others either determine the potential evapotranspiration (i.e. the rate of evapotranspiration, if the supply of water is unlimited), or evaporation equivalent to Glass A Pan evaporation, and then use the empirical relationships between these factors and the evapotranspiration. The method commonly adopted is to compute evapotranspiration from Glass A Pan evaporation either actually observed on evapori-meters or empirically determined from equations.

In India, the regional values of actual evapotranspiration have yet not been computed. However, evaporation data based on a network of about 80 stations equipped with wiremesh-covered standard Glass A Pan evapori-meters is now available for about 10 to 15 years. It has been observed that evaporation is lowest in Assam and the adjoining Himalayas and Bengal (the annual value being 150 cm) and the highest value is 387 cm at Jalgaon and Ahmedabad.

Evaporation is lowest (from 2 mm to 8 mm/day) during January and maximum during May (when it ranges from 4 mm to 20 mm/day). The monthly and annual potential evapotranspiration values have also been computed by applying the Penman formula for about 300 stations in India. The annual variation ranges from 140 cm to 180 cm over a large portion of the country. Values exceeding 200 cm, occur over western Rajasthan and parts of Saurashtra and Tamil Nadu. This is generally the highest during May (when it ranges from about 10 cm to 30 cm) over most parts of the country. In Rajasthan, it continues to be high in June also. The minimum is usually in December and January, when it ranges from about 3 to 15 cm per month.

Assessing Water Resources

Surface-water resources. The most reliable method to assess the surface-water resources is through an accurate and continual study of water flows through rivers and streams over a period of years. The runoff records of the various river systems in India are incomplete and therefore, the assessment of the water resources from actual gauged discharge records is not possible for the whole country. However, the rainfall records are available for a long period and attempts have, therefore, been made to roughly assess the water resources on the concept that the natural runoff (assuming that there is neither any upstream development from the surface- or groundwater resources nor any increment or decrement in the groundwater storage) is equal to the total volume of precipitation minus the volume of water lost into the atmosphere through evapotranspiration.

Dr A. N. Khosla, an eminent engineer, evolved the following relationships in the runoff as the function of rain fall and temperature, assuming that the temperature is the main determinant of evapotranspiration :

$$R_m = P_m - L_m \quad \text{....(1)}$$

$$L_m = 0.481\ T_m \quad \text{....(2)}$$

in which

R_m =monthly runoff, cm

P_m =monthly rainfall, cm

L_m =monthly evaporation losses, cm

T_m =mean monthly temperature, °C.

On the basis of this relationship, he estimated the mean annual natural runoff (normally referred to as the surface-water resources) of the six water-resources regions of the country as 167.23 mham. The region-wise distribution is given in Table.

	Name of Basin Catch-	*Average ment area (m ha)*	*Total annual precipi- tation (cm)*	*Mean precipi- tation (m ha m.)*	*Average tempera- ture (°C)*	*annual runoff (M ha m)*
1	2	3	4	5	6	7
1.	Indus Basin	35.40	56	19.82	12.6	7.94
2.	Ganga Basin	97.60	111	108.34	16.8	48.96
3.	Brahmaputra Basin	50.62	122	61.76	8.2	38.08
4.	West-flowing Rivers	49.16	122	59.98	25.5	31.06
5.	East-flowing Rivers	121.03	100	131.92	26.1	41.19
6.	Luni and Ghaggar Basin	16.80	29	4.87	26.2	-
	Total	370.61	549	386.69	-	167.23

Later, based on the specific analysis of the stream-flow data from 80 sources, carried out by the Central Water and Power Commission, the Irrigation Commission (1972) has worked out the water resources of the country at 178 mham.

This figure includes, besides surface runoff, groundwater runoff (effluent discharge) which is contributed by the groundwater storage as the base flow of the rivers (as shown later the present figure of groundwater runoff is roughly assessed at about 45 mham) as well as the sub-surface runoff contributed by rainfall and seepage from surface-water resources, but excludes whatever influent recharge (tank storage) that may have taken place from the flood flows.

The above estimate represents the average annual river flow. All this flow cannot be utilized owing to the highly variable character of the flow and other limitations imposed by the physiographic factors. The concentration of flow in a four-month period necessitates the construction of storage works, if the flood flows passing to the sea are to be utilized. However, suitable sites for the construction of storage dams are limited.

The utilizable quantum also depends on the availability of suitable sites for the diversion of water, the quality of water being suitable, the proximity of water to the land fit for cultivation and the dependability of flow. In India, the irrigation projects are usually designed for 75 per cent dependability, so that the designed quantity of water is available for at least 75 per cent of the year. In view of the limitations imposed by these considerations, the total utilizable flow has been estimated at 66.6 mham, by the Irrigation Commission (1972).

Region-wise details of the total available and the utilizable surface-water resources are given in Table.

S. No.	Name of Basin	Average annual runoff	Utilbisable flow
1.	Indus Basin	7.70	4.93
2.	Ganga Basin	51.00	18.50
3.	Brahmaputra Basin	54.00	1.23
4.	West-flowing Rivers	28.80	6.92
5.	East-flowing Rivers	34.80	33.80
6.	Luni and Ghaggar Basin	1.70	1.22
	Total	178.00	66.60

With further growth of irrigation, it is likely that subsurface flow contributing to the total annual runoff may appreciably increase, thus augmenting the present availability of surface-water resources. This increase, however, may be compensated for by the anticipated reduction in the groundwater runoff owing to the continued development of the groundwater resources. It is, therefore, visualized that the total surface-water resources may remain, more or less, steady at the same level as at present.

Ground Water : The assessment of groundwater resources is much more difficult, as it involves the evaluation of the various hydrologic components within the framework of a complex geological environment. There are as yet numerous gaps in the available information and the stage has not reached when any precise basin-wise assessment of the resource can be made except in highly generalized terms.

An attempt was made by the Central Working Group (constituted for working out the Fourth Five-Year Plan proposals for Minor Irrigation in 1968) to work out the rainfall contribution of groundwater recharge by classifying the geological formations in India into seven categories and assuming, on the basis of the best-informed judgement, different percentage factors for the groundwater recharge for these various categories.

The categories and the percentage factors assumed are indicated in table.

Rock type formation	*Assumed percentage rainfall contributing to ground water recharge*
1. Hard-rock formations and the Deccan traps	10
2. Consolidated rocks (sandstone)	5-10
3. River alluvia	15-20
4. Indo-Gangetic alluvium	20
5. Coastal alluvia	10-15
6. Western Rajasthan sand dune	2
7. Intermontane valleys	15-20

On the above basis, the average contribution to groundwater recharge in the country as a whole works out at 10 per cent of the total rainfall. Since then, some more reliable regional estimates (though still based on short-cut generalized methods) have been made by the state groundwater organizations, and it is felt that the figure of 10 per cent contribution, on the average, may be somewhat conservative. A figure of 12.5 per cent is considered to be more realistic.

Reckoning that the total annual volume of rainfall is about 400 mham, on the basis of a mean rainfall of 120 cm over the geographical area of 328 mham, the mean rainfall contribution to the groundwater recharge comes to 50 mham.

As mentioned before, the under-ground aquifers are supplemented from sources other than rainfall, seepage from canals, water-courses, field channels, ponds, tanks, return flow from irrigation in the fields and the influent recharge from rivers. As assessed later, the total volume of surface water presently utilized for irrigation is about 23.5 mham.

It is visualized that about 35 per cent of this total volume diverted for irrigation would pass as deep percolation to groundwater storage. Hence, the addition to the groundwater recharge on this account may be taken as about 8.2 mham. The total volume of groundwater presently used for irrigation is estimated at about 10.5 mham. Since in the case of groundwater used for irrigation, the loss in the long conveyance systems is not involved and the efficiency of irrigation is somewhat better, it may be more realistic to assume a figure of 25 per cent as the deep percolation contributing to groundwater recharge. The addition to groundwater recharge on this account will thus be 2.75 mham. The total contribution to groundwater recharge from these sources may thus be taken as 10.95 (say 11) mham.

Assuming that about 4 mham, may be the influent recharge from the flood flows, the total mean availability of groundwater resources at present may be reckoned as about 65 mham. It is roughly visualized to be accounted for as below:

1. Withdrawal use :	
(i) Irrigation	10.5 mham
(ii) Other uses	1.5 mham
Total	12.00 mham
2. Loss due to evapotranspiration from the water-table	8.00 mham
3. Groundwater rainfall (effluent recharge to rivers and streams)	45.00 mham
Total	65.00 mham

According to the assessment of the National Commission on Agriculture, the contribution from surface-water resources to the groundwater recharge is likely to increase to about 35 mham on the full development of irrigation—25 mham from seepage from canals, tanks, etc. and 10 mham as influent recharge (to be increased as a result of special measures for induced recharge). Adding the contribution of rainfall, the total ultimate groundwater resources may be of the order of 85 mham. Reckoning that about 50 mham will be lost as groundwater runoff and evapotranspiration, the ultimate usable ground water resources may be assessed at 35 mham. The above assessment does not include the additional supplies that may temporarily be obtained by the mining of water in certain regions.

Role of Water Resources and Usage : Water resources are utilized mainly for three purposes: irrigation, industrial use and domestic water supply, urban as well as rural. Irrigation is by far the major consumer of water resources.

Irrigational Pattern : The development of irrigation has received attention in the country since the beginning of the Indian civilization. Reference to wells and tanks has been made in the *Vedas*. Some of the existing tanks and canals were constructed centuries back. The British Government remodelled and constructed several major canals in the country during the nineteenth and twentieth centuries. Since the inception of the First Five-Year Plan, concerted efforts have been made to develop irrigation through major, medium and minor irrigation schemes (schemes costing more than its 5 crores are classified as major, those between Rs 25 lakhs and 5 crores as medium and those costing less than Rs 25 lakhs as minor).

It would appear that under minor irrigation, there has been a significant step-up in the development of groundwater, particularly since the end of the Third Plan. The factors responsible for this acceleration were: the realization of the value of groundwater as a source of assured irrigation, the rapid extension of electric power to villages ; the increased production and availability of diesel engines and diesel oil; the fast development in the extension of cooperative credit for minor irrigation ; and the indigenous production of drilling and other equipment required for the programme. Table gives the progressive increase in the number of wells and pumping sets:

Table gives the progressive development of irrigation since the beginning of the First Five-Year Plan.

Period	*Major-medium irrigation (surface)*	*Minor irrigation Surface*	*Ground*	*Total*	*Grand Total*
	(In million ha)				
Pre-Plan	9.7	6.4	6.5	12.9	22.6
End of the First Plan	11.0	6.5	7.6	14.1	25.1
End of the Second Plan	13.1	6.6	8.2	14.8	27.9
End of the Third Plan	15.2	6.8	10.2	17.0	32.2
End of the three-year period—1966-69	17.0	7.0	12.0	19.0	36.0
End of the Fourth Plan	19.4	7.5	16.0	23.5	42.9

Item	*1950-51*	*1960-61*	*1968-69*	*1973-74*
		('000 numbers)		
1. Dug wells	3,500	4,500	5,900	6,920
2. Shallow tube-wells (private)	3	20	250	810
3. Deep tube-wells (public)	3	8	15	20
4. Electrical pumping-sets	18	192	1,089	2,440
5. Diesel pumping-sets	65	230	722	1,753

With more and more area coming under irrigation, there has been a steady increase in the intensity of cropping as well as in the intensity of irrigated cropping. This increase is shown in Table.

Period	*Intensity of cropping*	*Intensity of irrigated cropping*
1950-51	111.12	108.13
1955-56	114.01	112.28
1960-61	114.63	113.36
1965-66	114.10	117.11
1968-69	116.07	122.07
1970-71	118.55	123.32

The total area irrigated from the surface-water sources in 1973-74 is estimated at 26.9 million ha. This area includes about 3 million ha in which the supplemental irrigation is provided by groundwater schemes. Reckoning an average total irrigation depth of 0.9 metre measured at the head of the source, for 23.9 million ha, irrigated exclusively from the surface-water schemes, the total volume of surface-water presently used (in 1973-74) comes to 21.51 mham.

Assuming that an irrigation depth of about 0.3 metre on the average will be provided by the groundwater schemes in the area under supplemental irrigation, the average depth provided by the surface-water schemes may be taken as 0.6 metre. The total volume utilized on this account will come to 1.8 mham. The present total utilization of surface-water for irrigation may then be taken as 23.31, say 23.50 mham.

For groundwater irrigation, as in this case conveyance is not involved and also the efficiency of irrigation is somewhat better, a reduced yard stick of 0.6 mham of irrigated area is considered to be realistic. On this basis, the total volume of groundwater utilized in 1973-74 for irrigating 16 million ha would work out at 96 mham.

In addition, it is roughly estimated that 9 mham metres of groundwater was used for providing supplemental irrigation to an area of about 3 million ha under the command of existing surface-water irrigation projects. The total ground water pumpage for irrigation in 1973-74 is thus assessed at 10.5 mham.

Other Uses : No scientific estimate is available of the present utilization of water resources for industrial and domestic purposes. It is, however, roughly reckoned that the total utilization for these purposes may be about 2.5 mham, about 1 mham from surface-water and 1.5 mham from groundwater.

Total Utilization : The following data give the picture that emerges in respect of the present status of utilization of the water resources :

	Total utilizable resources	*Present utilization*		
		Irrigation	*Other uses*	*Total*
			(In mham)	
1. Surface-Water	66.6	23.5	1.0	24.5
2. Ground-Water	35.0	10.5	1.5	12.0

Expectations in Future

Table gives the future perspective for irrigation development in the country based on the present state of knowledge about the usable water resources and the possibility of utilizing them:

	Ultimate gross irrigated area (mham)	*Average total depth of irrigation*	*Ultimate utilization of water resources for irrigation (mham)*	*Total usable water resources (mham)*	*Balance water resources for other uses (mham)*
(a) *Surface Irrigation :*					
(i) Exclusive of surface water	62	0.8	49.6	NA	NA
Supplemental irrigation*	10	0.47	4.7	NA	NA
Total	72**	-	54.3	66.6	12.3
(b) Ground Water :					
(i) Exclusive of groundwater	35	0.66	23.1	NA	NA
(ii) Supplemental irrigation	10*	0.33	3.3	NA	NA
Total	35		26.4	35	8.6

**This figure includes area partly irrigated from surface -water-works and partly from groundwater schemes.*

***It includes 57 m ha from major-medium works and 10 m ha from minor works.*

**Included under surface irrigation.*

In the above projections it has been assumed that with more emphasis on the lining of canals, distributaries and water-courses and improving the efficiency of irrigation, the average depth of irrigation in the case of surface irrigation will decrease from the present 0.9 m to 0.8 m in course of years. For groundwater irrigation, on the other hand, it is visualized that the present figure of 0.6 m may have a tendency to increase owing to the gradual extension of groundwater irrigation to higher-water-consuming crops, like paddy and sugarcane. Hence a figure of 0.66 m has been assumed in the context of the ultimate development from groundwater.

For the area under supplemental irrigation, it is assumed that an average irrigation depth of about 0.47 m will be provided from surface-water and 0.33 m from groundwater.

As regards the net area sown, the present figure which is about 142 m ha, the outer limit is considered to be 150 million ha in view of the paucity of additional land available for bringing under cultivation. The question is: What will be the realistic corresponding figure in the gross cropped area?

The data for the years 1955-56 to 1970-71 show the near-linear relationship between the area sown more than once and the gross irrigated area.

Further, if the past trend is assumed to continue, the ultimate figure of the areas taken more than once, when the gross irrigated area reaches the ultimate level of 107 m ha, will be 60 m ha This would mean that the ultimate total cropped area may be reckoned to be of the order of 210 m ha.

The past figure shows that the intensity of cropping has a tendency to be somewhat less than the intensity of irrigated cropping, although both of them increase with the growth of irrigation.

The ultimate intensity of cropping on the basis of the gross cropped area of 210 m ha and the net area sown covering 150 m ha will be 140. Assuming a somewhat higher intensity of about 150 per cent for irrigated cropping, the net irrigated area against the gross irrigated area of 107 m ha may be taken as about 70 million ha.

Table gives the overall perspective for the ultimate increases visualized in the areas under cultivation and irrigation.

	Item	*1973-74*	*Ultimate*
			(Million ha)
1.	Gross cropped area	169	210
2.	Net area sown	142	150
3.	Gross irrigated area	42.9	107
4.	Net irrigated area	34.8	70
5.	Intensity of cropping	119%	140%
6.	Itensity of irrigated cropping	123%	150%

The above figures give the basis on which the ultimate projections for agricultural production in the country may be made. The present trend shows that about 90 to 95 per cent of the ultimate potential for irrigation from groundwater may be reached by the year 2000. The full exploitation of the potential for irrigation from surface-water sources may extend to about the year 2025.

The water resources to be ultimately available for non-irrigation uses amount to 12.3 mham from surface-water and 8.6 mham from groundwater. Although the percentage of utilization of water resources for purposes other than irrigation is low at present, this is expected to rise appreciably in the future with increasing industrialization and power generation through thermal and nuclear plants. According to the National Commission on Agriculture, by the year 2025 the requirements under these sectors may amount to over 25 per cent of the available water resources and may increase further beyond this period. It is obvious that there will be constraints in meeting these requirements by the year 2025 from the normal resources.

The above perspective underlines the fact that in the face of anticipated population growth and technological advancement, the country's utilizable water resources are limited and there is no ground for complacency in the matter. There is utmost need for laying maximum emphasis on conservation and efficient utilization of the available water resources. To achieve this objective, the *laissez-faire* approach in water development should give way to integrated development based on carefully drawn-up plans for every agro-ecological region. Technological and management capabilities will have to be updated for this purpose. The degradation of the environment through pollution, which has so far been neglected, will need to be given serious consideration while making decisions relating to the future development. Above all, there is need that while the available water resources are utilized in the best manner possible, the development of technology for augmenting the water resources through new methods, as artificial recharge, recycling of water, desalinization of sea-water and weather modification should also be given constant attention.

3

Formation of Soil

Soil may be defined as a thin layer of earth's crust which serves as a natural medium for the growth of plants. It is the unconsolidated mineral matter that has been subjected to, and influenced by, genetic and environmental factors— parent material, climate, organisms and topography all acting over a period of time. Soil differs from the parent material in the morphological, physical, chemical and biological properties. Also, soils differ among themselves in some or all the properties, depending on the differences in the genetic and environmental factors.

Thus some soils are red, some are black; some are deep and some are shallow: some are coarse-textured and some are fine-textured. They serve as a reservoir of nutrients and water for crops, provide mechanical anchorage and favourable tilth. The components of soils are mineral material, organic matter, water and air, the proportions of which vary and which together form a system for plans growth : hence the need to study the soils in perspective.

Soil Formation with the Help of Materials : Rocks are the chief sources for the parent materials over which soils are developed. There are three main kinds of rocks:

(i) igneous rocks,

(ii) sedimentary rocks, and

(iii) metamorphic rocks.

Igneous Rocks : They are formed by the cooling, hardening and crystallizing of various kinds of lavas and differ widely in their chemical composition. They chiefly contain feldspars, maphic minerals and quartz. Bocks containing a high proportion of quartz (60-75%) are classified as acidic, whereas those containing less than 50% quartz are classified as basic. The common igneous rocks found in India are the granites (acidic) and basalts or the Deccan Trap (basic).

Sedimentary Rocks : They are derived from igneous rocks and are formed by the consolidation of fragmentary rock materials and the products of their decomposition deposited by water. The common sedimentary rocks are conglomerate, sandstone, shale and limestone. Alluvial, glacial and aeolian deposits form the unconsolidated sedimentary rocks.

Metamorphic Rocks : They are formed from the igneous or sedimentary rocks by the action of intense heat and high pressure or both resulting in considerable change in the texture and mineral composition. The common metamorphic rocks are gneiss from granite, quartzite from quartz or sandstone, marble from limestone and slate from shale.

The rocks vary greatly in chemical composition. Table gives the average chemical composition of four different kinds of rocks.

Percentage Chemical Composition of Rocks

Oxides	*Igneous rocks*	*Shales*	*Sandstones*	*Limestones*
SiO2	59.07	58.90	78.64	5.20
Al_2O_3	15.22	15.63	4.77	0.81
Fe_2O_3	3.10	4.07	1.08	0.54
MgO	3.45	2.47	1.17	7.92
FeO	3.71	2.48	0.32	-
CaO	5.10	3.15	5.51	42.74
Na_2O	3.71	1.32	0.45	0.05
K_2O	3.11	3.28	1.32	0.33
CO_2	2.67	5.03	41.70	-
P_2O_5	0.30	0.17	0.08	0.04
MnO	0.11	-	-	-
TiO_3	1.03	0.66	0.25	0.06
H_2O	1.30	3.72	1.33	0.56
Miscellaneous	0.79	1.48	0.07	0.05

Weathering refers to the physical and chemical disintegration and decomposition of rocks which are not in equilibrium under temperature, pressure and moisture conditions on the earth's surface. In the beginning, weathering precedes soil formation, more so in hard rocks. In other words, weathering creates the parent material over which soil formation takes place. Later, weathering, soil formation and development proceed simultaneously. The weathering may be physical or chemical.

Weathering Regarding Physical Nature : The principal agents of physical weathering are given below.

1. Temperature : The differential expansion and contraction of minerals in the rocks due to variations in temperature set up internal tensions, form weaker zones and

gradually break them apart. High temperatures accelerate the process of chemical weathering, especially in warm humid regions.

2. Water : Torrential rains and flowing waters dislocate the solid particles on the rocks and expose the inner portion to the agents of weathering. The dislodged particles are carried down and deposited elsewhere as alluvium. Similarly, the sea-waves wear off the rocks on the shore, and the glaciers in the high mountains exert an erosive and transporting influence on the rocks and their fragments.
3. Wind Flow : Wind exerts abrasive action, detaches the particles from the rocks and acts as a carrying agent. Sand-storms in the deserts and high winds on the sea-shore have both erosive and transportive action.
4. Forest and animals : Lichens and mosses growing on bare rocks cause their gradual disintegration. Grasses, shrubs and trees growing in rocks' crevices help to extend the cracks by the growth of their roots. The decomposition of litter and decayed matter accelerates the chemical weathering owing to the release of organic acids.

Weathering Regarding Chemical Nature : The chemical decomposition of rocks is brought about by solution, hydration, hydrolysis, carbonation, oxidation and reduction. Chemical weathering taking place in the lower layers may be termed geo-chemical weathering, whereas that taking place at the surface and under the surface may be termed pedo-chemical weathering.

1. Solution Factor : The solvent action of water is an important means of weathering rocks containing soluble salts; e.g. gypsum and limestone. The solvent action is increased in the presence of carbon dioxide and organic-

acids released during the decomposition of organic matter. Salts of sodium, potassium, calcium and magnesium are readily removed in solution.

2. Hydrational Factor : Hydration implies the association of water molecules with minerals. It provides a bridge or entry for the hydroniumions to attack the structure.
3. Hydrolysis : It is the most important process of chemical weathering. During hydrolysis, the hydronium ions (from water) attack the weatherable minerals and alter them completely, or modify them drastically. In the process, hydroxides of potassium, iron, magnesium, calcium, etc. are formed.
4. Carbonisational Factor : The hydroxides produced during hydrolysis react with the dissolved carbon dioxide to form corresponding carbonates which may either leach out or accumulate according to drainage and weather conditions.
5. Oxidational and Reduction Factor : Oxidation is an important reaction in well-aerated rocks and soil material where oxygen supply is high and biological demand is low. For example, the oxidation of iron is a disintegrating weathering process in minerals containing ferrous iron as part of their structure and the ferric' iron due to its size and charge breaks up the mineral.

During reduction (the reverse of oxidation), when the supply of oxygen is low but the biological demand is high, the ferric iron is reduced to the ferrous iron which, being more mobile, may leach downwards. In this manner, oxidation and reduction are responsible for weathering minerals.

The mineral weathering combined with the associated physical and chemical phenomena constitutes the processes of soil formation. Soil formation is a complex of events. One process is distinguished from another on the basis of the chemical nature of events. The processes of soil formation include: (i) the addition of organic and mineral materials, (ii) the loss of these materials from the soil, (iii) the translocation of materials from one point to another within the soil column, and (iv) the transformation of mineral and organic substances within the soil.

Weathering and soil formation showing the influence of genetic and environmental factors lead to soil development. The kind and intensity of weathering and processes of soil formation indicate the degree of soil development, as judged from the examination of a soil profile. Successive stages of soil development may be called infancy, youth, maturity and old age, introducing the time factor in them. For attaining maturity and old age, the positional stability of soil over a relatively long period to show the influence of intensive processes is essential.

In the steep Himalayan region, landslides and avalanches frequently move large masses of soft rocks and their weathered products. Similarly in deserts, the region is subject to fresh wind-borne deposits. In these situations, therefore, the soils may never reach the stage of maturity or old age. The soils in the active flood plains which receive frequent fresh deposits will also continue to remain youthful in terms of soil development. Differences in relief play an important part in soil formation by draining away water and finer fractions from high levels to depressions and lowlands. Thus the soils formed at higher levels under low moisture conditions are coarse-grained as compared with those of lowlands which are finer in texture. Though soil differences in the initial period arise from differences of parent materials, yet in a large measure, they are profoundly modified by other factors of soil formation, namely climate, organisms, topography and time.

Profile of Soil : It is a vertical section of the soil through all its horizons and it extends up to the parent materials. A study of soil profile is important both from the standpoint of soil formation and soil development (pedology) and crop husbandry (edaphology). In deep soils the soil profile may be studied up to one metre and a quarter and in others up to the parent material. The layers (horizons) in the soil profile which vary in thickness may be distinguished from the morphological characteristics which include colour, texture, structure, etc. Generally, the profile consists of three mineral horizons 'A', 'B' and 'C.

The 'A' horizon may consist of sub-horizons richer in organic matter intimately mixed with mineral matter 'A', the horizon that has lost clay, iron or aluminium with the resultant concentration of quartz (A_2—eluvial horizon) or an intermediate or transitional horizon.

The 'B' horizon is below the 'A' horizon showing dominant features of the concentration of clay, iron, aluminium of humus alone or in combination.

The 'O' horizon excludes the bedrock from which 'A', and 'B' horizons are presumed to have been formed. The 'C' horizon by itself is relatively little affected by pedogenic processes. It is the layer of consolidated bedrock which, upon weathering, may be presumed to have given rise to the 'C horizon above.

Not all the profiles show the sequence mentioned above. Profiles developed *in situ* under intense pedogenic processes over a sufficient period show the presence of the horizons 'A', 'B' and 'C. Since the formation of soil and the development of profile are dependent on the genetic and environmental factors which vary considerably within and between regions, variations in horizonation are frequent and common. Thus, for example, the soils developed in a recent flood plain may have only 'AC profile without any 'A_2', whereas those in the red and laterite soils area may have 'A'_1', 'B'_2 and 'C. In some forest soils, organic layers may also be found on the surface above 'A'.

It will be beyond the scope of this brief chapter to deal at length the nomenclature, implication and significance of horizons. It may be mentioned that a study of morphology of the soil profile gives a clear expression of the influence of the pedological processes involved.

A study of the soil profile is important from crop husbandry point of view, since it reveals the surface and the subsurface characteristics and qualities, namely depth, texture, structure, drainage conditions and soil-moisture relationships, which directly affect plant growth. Hence the soil profile is taken as a unit of study which helps the investigators both to classify the soils and to understand soil-moisture-plant relationships. A study of the soil profile in the field, therefore, furnishes a base which has to be supplemented by physical, chemical and biological properties of the soils briefly discussed separately.

Physical Characteristics

The physical properties of a soil largely determine the manner in which it can be used. Properties, e.g. the water-holding capacity, permeability to water, aeration, plasticity and nutrient-supplying ability, are influenced by the size, proportion, arrangement and mineral composition of the soil particles. The proportion of the four major components of the soils—inorganic or mineral particles, organic material, water and air—vary greatly from place to place and with depth. The amount of water and air in a soil fluctuates from season to season, but the proportion of primary solid components of the soil, however remains unchanged.

Particle shape and size. The soil particles vary in shape from spherical to angular. They differ in size from gravel and sand to fine clay. The effective size limits adopted are given in Table:

Particle Sizes—USDA and International Classification Systems

	International or Atterberb system	*USDA system*
	(Diameter in mm)	
Gravel	2 and more	2 and more
Very coarse sand	2-1	
Coarse sand	1.0-0.5	2.0-0.2
Medium sand	0.5-0.25	
Fine sand	0.25-0.1	0.2-0.02
Very fine sand	0.10-0.05	
Silt	0.05-.002	0.02-0.002
Clay	less than 0.002	less than 0.002

While the International System is commonly used to designate textures, the US system is necessary for classifying soils according to Soil Taxonomy which is being adopted in the soil-survey organizations all over the country.

Textural Classes : The varying proportions of particles of different size groups in a soil constitute what is termed as the textural class.

The principal textural classes are : Clay sandy clay, silty clay, clay loam, sandy clay loam, silty clay loam, loam, sandy loam, silt loam, sand, loamy sand and silt.

The physical properties and the chemical composition of the small and large particles differ greatly. The coarse fraction, gravel and sand which are mainly composed of rock fragments or primary minerals act as individual particles. They have low specific surface and are relatively non-reactive . They do not hold large amounts of water or nutrients. Owing to large voids between them, they transmit air and water easily.

The silt particles are intermediate between sands and clays. Mineralogically, the particles of silt are similar to those of sand, as they are largely composed of primary minerals. They are more reactive than sands because of the higher specific surface.

The clay fraction controls most of the important properties of the soils. They are chiefly composed of secondary minerals-crystal line alumino silicates. They have high specific surface and are most reactive. They have high capacity to retain water and nutrients.

The textural classes differ not only in the particle size analysis, but also in their bearing on some of the important factors affecting plant growth, such as (i) the movement and availability of water, (ii) aeration, (iii) workability, and (iv) the content of plant nutrients.

Sandy soils are very permeable and well-drained but are less water-retentive and hence need more frequent irrigation for successful crop growth than fine-textured soils. The clayey soils can hold more moisture, but they have high wilting percentage. The rate of water intake of these soils is low. They are subject to water-logging, resulting in poor aeration and workability. The moderately fine-textured soils, e.g. loams, clay loams or silt loams are by far the excellent soils for crop growth, since they have the advantages of both sands and clays.

Soil Colour : Colour gives a ready clue to soil conditions and some important properties. It is due either to mineral matter or organic matter and mostly to both. Red, yellow or brown colours are mostly related to the different degrees of oxidation, hydration and diffusion of iron oxides in the soils. Dark colours of soils are associated with one or a combination of several factors, including impeded drainage conditions, content and state of decomposition of organic matter, the presence of titaniferous magnetite, etc.

Uniformity in nomenclature of colours is possible by comparing the soils with charts containing standard colours.

One such chart is the Munsell Colour Chart. It consists of coloured chips arranged according to hue, value and chroma, the three simple variables that combine to give the colour. The hue refers to the dominant spectral colour, the value to relative lightness of colour and chroma to relative purity. Density. Soils having larger particles are usually heavier in weight per unit volume than those with smaller particles. True density of a soil is based on the individual densities of soil constituents and according to their proportionate contribution. The bulk density or a apparent density is the weight per unit volume of dry soil as a whole, i.e. particle and pore space, and hence it is lower than the true density. The relationship between the true density (T) and the apparent density (A) and the pore space (P) is as follows :

$$P\% = \frac{T-A}{T} \times 100$$

In most mineral soils the true density varies within narrow limits of about 2.5 to 2.7 and the apparent density between 1.4 and 1.8.

Pore Space in the Soil : The pore space of soil is the portion occupied by air and water, and it is determined largely by structural conditions. Sands have low pore space of about 30 per cent, whereas clays may have as much as 50-60 per cent. Good loams may have pore space of 40-50 per cent. Although clays possess greater total porosity than the sands, the pore spaces in the latter, being individually larger, are more conducive to good drainage and aeration.

Plasticity and Cohesion of the Soil : Plasticity is the property that enables a moist soil to change shape on the application of force and retain this shape even when the force is withdrawn. On this basis, sandy soils may be considered to be non-plastic and clayey soils to be plastic. Cohesion is the tendency of the particles to stick to one another. Plastic soils are cohesive. Plasticity and cohesion reflect the soil consistency and workability of the soils.

Temperature and Heat : Soil temperature is one of the important factors that control the microbiological activity and all the processes involved in the growth of plants. Heat is necessary for seed germination, root growth and other biological activities. The temperature needed for germination and root growth varies with crops and varieties. Crops, e.g. wheat, barley and peas, grown in India during winter germinate at relatively low temperature as compared with maize, and those at which groundnut or cotton germinate. Microbiological activities are retarded by low soil temperature. As a result, the nitrification processes in the soil are slowed down and plant nutrition and growth adversely affected.

The temperature of the soil is influenced by its colour, composition, slope, aspect and water content. Dark soils absorb more heat than those of lighter hues. Generally, sandy soils absorb heat during the day and lose it during night more quickly than the fine-textured soils, because the latter retain more water and the specific heat of water is 4 to 5 times more than that of the soil particles. Soil temperature is considered so important that in the soil taxonomy there is a provision to use it as a differentiating criterion at the family level of categorisation.

Air in the Soil : It is well known that restricted soil aeration adversely affects root development, processes of respiration and other essential biological processes. It is, therefore, important to know the content of soil air and its composition. It has already been mentioned that, depending on its texture, the soil may have pore space of 30-60 per cent. The pore space, not filled by water, is occupied by air. Under moist field conditions, the non-capillary pore space (macropores) generally constitutes the air space, the capillary pore space (micropores) being occupied by water. Ordinarily, the occupation of nearly one-third of the pore space in the soil by air and two-thirds of it by water constitutes the most favourable condition for plant growth.

The composition of soil air reflects a dynamic balance between two competing processes. The consumption of oxygen and the liberation of carbon dioxide by plant roots and the soil organisms tend to increase the difference in the composition of the soil air and the atmosphere above the soil surface. Gaseous diffusion tends to reduce the difference in composition.

Like atmospheric air, the soil air is composed largely of nitrogen and oxygen, but differs from the former in containing more moisture, more carbon dioxide and a little less oxygen. The actual content of carbon dioxide varies greatly during the crop season because of the higher rates of crop growth and biological activities in the soil.

The composition of soil air under different conditions is given in Table.

Percentage Composition of Surface-Soil Air of Different Soils*

Kind of Soil	*Percentage composition*		
	Carbon dioxide	*Oxygen*	*Nitrogen*
Forest soil	0.87	19.61	79.52
Clay soil	0.66	19.61	79.35
Asparagus bed not manured for one year	0.74	19.02	80.24
Asparagus bed freshly manured	1.54	18.80	79.66
Vegetable mould compost	3.64	16.45	79.91

Contd...

Kind of Soil	*Percentage composition* *Carbon dioxide*	*Oxygen*	*Nitrogen*
Swamp rice land	4.23	0.31	84.35
	to	to	to
	4.69	0.99	86.60**
Vicinity of sunnhemp roots	12.12	2.23	79.77
	to	to	to
	16.99	4.67	81.63
Vicinity of maize roots	3.34	7.25	74.47
	to	to	to
	12.30	13.82	81.59

Compiled from different sources. Other gases formed 8.41 to 11.09 per cent of air in this soil.

Besides free air, the soil contains air held by colloids or that dissolved in soil solution. This air contains about 90 per cent carbon dioxide, about 10 per cent nitrogen and a trace of oxygen.

Water Capacity of Soil : Water has perhaps the greatest influence on the growth and yield of a crop. It is needed in much larger quantity than that of any other substance that contributes to growth and yield. Water serves the following functions in relation to plant life.

(i) It is an essential part of plant food. It constitutes nearly 90 per cent of plant tissues.

(ii) It serves as a solvent and carrier of plant nutrients.

(iii) It maintains cell turgidity and regulates temperature. Water is held in the soil in the following forms :

1. Water Due to Hydroscopic : It occurs as a thin film (4-5 millimicron) and is held tenaciously with a

tension of 31 atmospheres or more. It is not available to plants.

2. Water Due to Capillary : It forms a continuous film around soil particles (outside the film of hygroscopic water) and in the micropore spaces. It is held by surface tension. Capillary water is held at a tension ranging from $\frac{1}{3}$ to 31 atmospheres.
3. Water Due to Gravitation : It is free water held at a tension below $\frac{1}{3}$ atmosphere. It saturates the soil and percolates downwards under the influence of gravity.

The maximum amount of capillary water remaining in the soil after the removal of gravitational water is called its field capacity. In most cases, it represents the water held at a tension of $\frac{1}{3}$ atmosphere. It is generally recognized that capillary water held at tensions greater than 15 atmospheres is not available to plants. At this point of soil moisture, the plant wilts permanently and, hence, the percentage moisture at 15 atmospheres is called its wilting point or wilting percentage. It follows, then, that the moisture held at tensions between $\frac{1}{3}$ atmosphere (field capacity) and 15 atmospheres (wilting-point) is the water available to plants to the maximum extent. Thus capillary water is the major source of water used by plants.

Capillary water is capable of movement upwards, downwards or laterally, the movement taking place from the thicker part of the film to the thinner part. Similarly, as the capillary-water zone moves farther from the water-table below, the rate of movement becomes less and the suction needed to draw up water increases. Hence, in general, a water-table lying more than 2 to 3 metres below the root-zone is not of much use to crops.

Crops differ greatly in their water requirement according

to growth characteristics, climate and water-supply. Water requirement of a crop includes the evapotranspiration needs, the water needed for metabolic activities for leaching and other unavoidable losses. It is the water needed for raising a crop in a given period and is expressed as depth of water in millimetres. Generally, the water requirement of a crop is related to the potential evaporation during the growth of the crop.

Structure of Soil : Soil structure refers to the arrangement of soil particles, both primary and secondary. Soil structure is recognized as one of the most important properties of the soil mass, since it influences aeration, permeability, water capacity, etc. In the field, the structure is described in terms of (i) type-referring to shape and arrangement, (ii) class-referring to size, and (iii) grade-referring to the degree of aggregation.

Different Types : There are 4 types of primary structures:

1. Platy Structure : With particles arranged around a plane, generally horizontal.
2. Prism-like Structure : With particles arranged around a vertical axis and bounded by relatively flat vertical surfaces. This type includes prismatic and columnar types.
3. Block-like Structure : With particles arranged around a point and bounded by flat or rounded surfaces which are the casts or the moulds formed by the faces of the surrounding peds. This type includes angular and subangular types.
4. Spheroidal Structure : Particles arranged around a point and bounded by curved or very irregular surfaces that are not accommodated in the adjoining aggregates. This type includes granular and crumb types.

Various Classes : Five size classes are recognized in each of the primary types. They are very fine, fine medium, coarse and very coarse. The actual size for classes in each type varies.

Different Grade : The grade of structure, representing the degree of aggregation, is determined in the field mainly by noting the durability of the aggregates and the proportions between aggregated and unaggregated material that results when the aggregates are displaced or gently crushed. Grades are termed structureless (massive if coherent, and single-grained if non-coherent), weak, moderate, strong and very strong, depending on the stability of aggregates when disturbed.

Structure Affecting Factors: Structure is primarily influenced by texture. The beneficial effects of organic matter on the structure and working of the soils are well known. Soil structure is also influenced by :

1. Management of Soil : The cutting action of ploughs or other tillage implements breaks up the soil mass and may have favourable or adverse effects on the structure, depending on whether the soil is worked under optimum moisture conditions or not. A good soil management, with a proper system of crop rotation has the effect of maintaining the soil in a good state of aggregation.
2. Absorbing Caution : Sodium and potassium ions on the clay complex have a tendency to disperse the soil; calcium, on the other hand, has favourable effects on the aggregation. Similarly, the presence of soluble salts favours Flocculation.
3. Different Micro-organisms : The filamentous growth of soil fungi and the microbial decomposition products of organic matter have a binding effect on soil particles, thus favouring aggregation. Similarly, the excrements of earth worms and the burrowing activities of insects and other creatures cause appreciable changes in the soil structure.
4. Soil Moisture Variation : Variations of soil moisture due to drying and wetting influence the structure. The drying

of soil, in general, forms cracks and big clods. Poorly drained soils, with excess of moisture, usually have an unfavourable structure.

The structure of cultivated soils have a very important bearing on their drainage, ease of tillage, root penetration and resistance to erosion and ultimately their general productivity. From this point of view, the crumb and granular structure (spheroidal) is considered very favourable to plant growth.

The structure and physical qualities of soils can be improved by adopting a suitable system of soil management, including legumes in the rotation system, green-manuring, and regularly supplying the soil with organic manure.

Mineral composition of the soil separates. Owing to a varying rate of wearing down of different minerals, the more easily decomposing feldspars predominate in the relatively fine separates, whereas the resistant minerals, e.g. quartz, are more abundant in coarser separated. The separates, therefore, show differences in chemical composition also. As indicated earlier, the clay which has a high specific surface is more reactive than either silt or sand.

The clay fraction. In the process of decomposition, relatively stable new minerals, are formed from the products of weathering, which constitute largely the clay fraction. The clay particles formed in this manner are crystalline and, as shown by X-ray diffraction and petrographic techniques are composed of sheets of hydrated alumina and silica linked by oxygen atoms. Because of their being very small they are highly reactive and form the seat of ion exchange in soil and this controls and regulates adsorption retention, and release of many plant nutrients, such as potassium, calcium, magnesium and phosphorus. The precise nature of clay properties depends on the type of minerals that predominantly compose the clay. These minerals fall into the following three main groups :

1. The Kaolinite : A unit kaolinite crystal lattice consists of one sheet of silica and one sheet of alumina and is, hence, called l:l-layer silicate. The two sheets are held together by mutually shared oxygen atoms. These units, in turn, are tenaciously held together through oxygen hydroxyl linkages and, consequently no expansion occurs when the mineral is wetted. It has a low specific surface, and a low cation-exchange capacity. Similarly, plasticity, cohesion, shrinkage and the swelling properties of kaolinite are low.
2. The Montmorillonite : A unit montmorillonite crystal lattice consists of 2 sheets of silica and 1 sheet of alumina (2:1-layer silicate) held together by mutually shared oxygen atoms. Such units, however, are held together through weak oxygen-oxygen linkages. There is some isomorphic substitution of iron or magnesium in the alumina sheet; montmorillonite crystals can expand, and owing to this expansion, cations and water molecules are able to move in between the crystal units. Montmorillonite has a high specific surface and cation-exchange capacity. Similarly, plasticity, cohesion, shrinkage and swelling properties are high.
3. The Illite : Illite has the same general structural organisation as montmorillonite, except in respect of the linkages between the crystal units. Here, about 16% of silica in the silica sheet is replaced by aluminium and potassium atoms and they supply the additional connecting linkages between the crystal units, due to which illite shows a lower expansion capacity. It has properties intermediate between kaolinite and montmorillonite.

Colloids of Soil : The most active portions of the soils are those which are in the colloidal state. The colloidal state implies a two-phase system in which the material called the dispersed phase (fine clay and humus) is dispersed in the dispersed medium (water). In soils, the mineral and organic colloids exist in intimate and heterogeneous admixture.

The mineral colloid is present almost exclusively as the clay of various kinds, whereas the organic colloid is present as humus. It is generally supposed that clay particles less than 1 micron in diameter possess colloidal properties and these properties increase with a decrease in the size of the particles. The most distinctive colloidal properties are (i) the large specific surface or interface, and (ii) the capacity to hold solids, gases, salts and ions.

Depending on their nature, the soil colloids determine many of the physical and chemical properties. The colloidal material may occur as a thin gelatinous film around coarser particles, or it may occupy a considerable part of the space between the larger particles, thus serving as a binding material.

Soil colloids, especially the mineral colloids, exhibit cohesive and adhesive properties. Soils with a high amount of colloidal clay compete with plant roots for water and mineral nutrients, especially at lower levels of their availability. Generally speaking, the soil colloids have a high exchange capacity, which increases with silica sesquioxide ratio.

Chemical Characteristics

Mineral matter and its composition. The principal minerals occurring in the earth's crust are :

1. Feldspars : Feldspars are anhydrous alumino silicates of potassium, sodium and calcium.
2. Quartz : Quartz or silica is silicon dioxide found in crystalline rocks. It also occurs as free silica or sand.
3. Mica : Mica is a layer of alumino silicate of potassium; iron, magnesium and sodium may be present in varying proportions.

4. Limestone : It is largely calcium carbonate; with magnesium carbonate it occurs as dolomite.
5. Hornblende and augite : They are the ferro magnesium minerals consisting of silicates of calcium, magnesium, iron and sodium in varying proportions.
6. Olivine and serpentine : Olivine is an olive-green ferro-magnesium silicate. Serpentine is a hydrated silicate of magnesium.
7. Clays : Clays are secondary minerals. They are hydrated alumino silicates.
8. Other minerals : Other minerals occurring in the soil include tourmaline—a boro alumino silicate with alkali metals and iron or magnesium; rutile-titanium oxide, zircon-zirconium silicate, glauconite-hydrated silicate of iron and potassium, apatite-calcium phosphate, sulphur-bearing minerals, etc. Allophane and the hydrous oxides of aluminium, iron and titanium also commonly occur in soils and more so in highly weathered and leached soils.

Mineral	*Approximate percentage*
Feldspars	48
Quartz	36
Micas	10
Limestone and dolomite	2
Hornblende and augite	1
Olivine and serpentine	1
Clays	1
Other minerals	1

In Organic Constituents : From what has been stated in the preceding section, it is evident that compounds of silicon, aluminium, calcium, magnesium, iron, potassium and sodium form the principal chemical constituents of the mineral matter in the soils. Besides, the soil contains small amounts of a large number of other mineral elements, e.g. phosphorus, boron, manganese, copper, sulphur, zinc and cobalt. Thus the soil supplies all the following essential mineral elements required by the plants :

1. Macro Constituents : Phosphorus, potassium, calcium, magnesium and sulphur.
2. Micro Constituents : Iron, manganese, zinc, copper, molybdenum, boron and chlorine.

(Plants obtain carbon, hydrogen and oxygen from carbon dioxide and water. Soil is also the source of nitrogen for plants, but the ultimate source of nitrogen is traceable to atmosphere from where nitrogen fixation takes place physico-chemically and biologically).

The total amount of elements contained in the soils depends partly on the nature of the parent material from which they are formed and partly on their age and extent to which soluble products have been leached down.

The chemical composition of different horizons of a soil also shows a good deal of variation. Generally speaking, the 'A' horizon is richer in insoluble compounds than the 'B' horizon. Usually, some of the elements that are commonly leached out are the ones that are also required by plants. Tables give the percentage inorganic composition of eight representative Indian soils and of the successive horizons of the soil.

Exchange of Ion : It is a reversible process by which cations and anions are exchanged between solid and liquid phases and between solid phases which are in close contact with each other. The exchange of cations and anions is termed cation exchange and anion exchange, respectively. Ion exchange is the most important of all the processes occurring in a soil. Soil colloids, as stated previously, are the seats of ion exchange.

Chemical and physical processes connected with ion exchange include the weathering of minerals, nutrient absorption by plants, the swelling and shrinkage of clay, the leaching of soluble salts, etc.

* Cations are positively charged ions; calcium, hydrogen, magnesium, potassium ions, etc. are cations. Anions are negatively charged ions; nitrates, sulphates, chlorides and phosphates are anions.

Percentage Inorganic Composition of Representative Indian Soils

Type	State	SiO_2	Al_2O_3	Fe_2O_3	MnO	CaO	MgO	K_2O	Na_2O	P_2O_5
Sandy	Rajasthan	91.72	2.92	2.36	-	0.35	0.78	0.33	0.08	0.08
Sandy loam	Uttar Pradesh	84.84	5.30	4.52	-	0 91	0.52	0.16	0.03	0.10
Sandy loam	Punjab	-	-	10.89	-	0.06	0.29	0.71	-	0.34
Loam	Tamil Nadu	82.11	5.19	5.70	2.12	1.10	1.27	0.24	0.86	0.06
Loam	Uttar Pradesh	86.98	4.10	3.52	-	0.42	1.29	0.56	-	0.06
Silt loam	Assam	86.49	9.16	7.00	0.21	0.14	0.90	1.85	-	0.24
Clay	Tamil Nadu	65.16	13.76	9.27	0.25	2.18	2.47	0.14	0.01	Tr
Clay	Maharashtra	63.80	Not available		-	1.09	1.60	1.60	-	0.23

Percentages Chemical Composition of Different Horizons of Red Soil from the Nilgiris District, Tamil Nadu

Hori-zons	Depth (cm) (soluble)	SiO_2	Fe_2O_3+Al_2O_3	CaO	MgO	K_2O	N
1. 0-22	0.53	32.60	0.140	0.08	0.656	0.138	0.154
2. 22-75	0.58	30.22	0.084	0.02	0.667	0.138	0.109
3. 75-112	0.57	37.15	0.056	0.03	0.568	0.149	0.050

The capacity of soils to adsorb and exchange cations and anions varies greatly with the nature and amount of clay, and the organic matter. The cation-exchange capacity (CEC) is defined as the amount of a cation species bound at pH 7 (neutral pH) and is expressed as milliequivalents (m.e.) per 100 grammes. The CEC values of reference clay minerals are:

Kaolinite	3-10 m.e./100 g
Illite	10-30 m.e./100 g
Montmorillonite	80-150 m.e./100 g

Organic colloids may have CEC of 200 m.e. per 100 g or more.

Saturation of Base : The portion of CEC accounted for by the basic ions (calcium, magnesium, potassium, sodium) and expressed as percentage of the CEC is the base saturation percentage. A soil saturated with calcium and magnesium is considered normal and fertile. If a soil has more than 15% exchangeable sodium, it is considered to be an alkali soil. On the other hand, if the soils are base-unsaturated, i.e. the proportion of exchangeable hydrogen is more, the soil tends to be acidic.

Acid, saline and alkali soils are separately dealt with. Similarly, other items, viz. pH and conductivity, are discussed in the section 'Soil-Testing.'

Organic Elements

Organic matter, though forming a small part of mineral soils, plays a vital role in the productivity and conditioning of soils. It serves as source of food for soil bacteria and fungi which are responsible for converting complex organic materials into simple substances readily used by the plants. The intermediate products of decomposition of fresh organic matter help to improve the physical condition of the soil. The addition of organic matter also improves the working quality or friability of the soil. In association with clay and calcium, it helps to form the aggregates of soil particles to produce the 'crumb structure'. Applied in adequate quantity, organic matter serves as a mulch.

Humus in the Soil : The organic matter in the soil consists largely of plant remains, the residues of soil micro-organisms feeding on them and several products of their decomposition. In consonance with its origin, organic matter contains most of the mineral elements found in the plants. The plant remains may occur in a recognizable form, but more commonly they are found as a fairly stable dark, amorphous complex colloidal substance called humus. Chemically, humus represents a mixture of decomposed or altered products of carbohydrates, proteins, fats, resins, wax and other similar substances. These complex compounds are gradually decomposed by soil organisms into simple mineral salts, carbon dioxide, water, organic acids, ammonia, methane and free nitrogen, depending upon the initial composition of the organic matter. The average composition of humus is as follows;

	Percentage
Carbon	50
Oxygen	35
Nitrogen	5
Hydrogen	5
Ash (containing phosphorus, potassium, sulphur and other elements)	5

Humus is soluble in water only to a very little extent, but owing to high specific surface, it can absorb gases as well as water. Lime has a precipitating action on humus. Humus also exhibits exchange properties.

Important Content of Organic Factor : The content of organic matter varies with the kind of the soil, vegetation, climate cultivation, manorial and rotation practices, and biological activities. The vegetation determines the quality and quantity of organic material added each year, whereas the climate, particularly the temperature and moisture conditions, determine the rate of decomposition. In the Indian soils, the content of organic matter is generally low because of the high rate of decomposition under tropical and subtropical climate. Except in a few localized areas in the hilly and high-altitude regions, the organic matter in most of the cultivated soils rarely exceeds 1 per cent. The following figures illustrate the variation of the organic carbon of the surface layers of 3 important soil groups of India.

Soil group	*Percentage organic carbon*
Deep black soil	0.34-0.77
Red and laterite soil	0.68-6.53
Alluvial soil	0.28-1.10

The movement and distribution of humus in the soil depends on the pedogenic processes. Normally, the amount of organic matter and humus decreases from the surface downwards. Usually, in the forest soils, owing to continual leaf fall, the amount of organic matter is high. Under acidic conditions, humus becomes dispersed and is carried down into the subsoil where it may form part of the hard accumulated layer of mineral and organic colloidal matter.

Carbon-nitrogen Combination : Nitrogen plays a major role in the growth and reproductive processes of plants. The organic combination of nitrogen, with organic matter in the soil constitutes the principal store-house from which nitrogen is slowly made available to the crops. The relation of organic matter to nitrogen is of particular interest. Investigations have shown that the ratios among nitrogen, carbon and organic matter in the cultivated soils are fairly constant. From the average composition of humus given already, it is evident that the carbon : nitrogen ratio of organic matter is 10 to 1.

The general level of carbon and nitrogen in most of the Indian soils is low and the carbon : nitrogen ratios fluctuate widely from 5 to 25, with the average value around 14.

Various Soils and Micro-organism

Raw organic matter in the soil is not directly used by the plants as food. It must be broken down first into humus and then into simpler products before it can be so utilized. This work is done by different kinds of micro-organisms, which inhabit the soil by billions. The decomposition of organic matter primarily forms part of the feeding and growth processes of these microscopic plants and animals. Sugars, starches and proteins are broken down first; then cellulose and fatty substances (lipoids); and lastly, lignin or woody substances. When organic matter becomes chemically and biologically inert, the soil becomes infertile, despite its presence.

However, not all the soil organisms are beneficial. There are certain bacteria which under anaerobic conditions of water-logged soils cause denitrification, releasing free nitrogen which gets lost into the air. Some others cause plant diseases. It has also been claimed that intermediate products of bacterial decomposition of soil organic matter or certain organic products synthesized in the soil by some micro-organisms have important growth-promoting or plant-protecting (i.e. antibiotic) properties.

Soil Organism and Its Kinds : The soil micro-organisms have been classified as follows :

Microflora: (i) Bacteria, (ii) actinomycetes, (iii) fungi, and (iv) algae.

Microfauna: (i) Protozoa, and (ii) nematodes.

Besides microfauna, the soil harbours a large number of worms and insects of different kinds and sizes.

Population Per Hectare : The soil organisms vary in number from a few per hectare to many millions per gramme of soil. The density of population is determined by food-supply, moisture, temperature, physical condition and the reaction of the soil. In more or less neutral soils, bacteria dominate over other types of microscopic life. If the soil is acidic and rich in organic matter, fungi predominate. Algae abound on or near the surface of the soil in constantly moist and shady situations.

Under favourable conditions, such as those obtaining after the ploughing in of a green-manure crop, the bacteria multiply enormously. On the other hand, in sandy desert soils or under water-logged conditions, they are very scarce.

As among the soil micro flora, it is estimated that generally bacteria form about 90 per cent of the total population, actinomycetes about 9 per cent, and the fungi and algae together about 1 per cent. Among the soil fauna, protozoa are the most abundant, followed by nematodes, worms and insects.

Bacterial Consequence : The soil micro flora typically produce ammonia from organic compounds when they set free more of nitrogen from the organic matter on which they are living than what they can assimilate and convert into their own protoplasms. Ammonia, so released, is converted into nitrites by one group of organisms called *Nitrosomonas* and nitrites are further converted into nitrates by another group of organisms called *Nitrobacter.*

The process of conversion of ammonia into nitrate is called nitrification. The nitrate-forming bacteria are generally confined to the top 25 to 30 cm of the soil, where the content of organic matter is also more. These organisms are most active between 25°C and 38°C and under favourable conditions of tillage, aeration, neutral soil reaction and moisture content at field capacity. The bacteria cease or reduce their activity when the pH value of the soil falls below 5. They, however, become active again when the pH is raised to about 6 by liming. The large supplies of carbon in the raw organic matter favour an enormous multiplication of bacteria which use up the nitrates released by the nitrifying bacteria. A similar effect may also be produced sometimes by heavy green-manuring. This problem can be dealt with either by adding sufficient fertilizer nitrogen to the organic material to provide for the nitrogen needs of the micro-organisms and the main crop or by incorporating the organic matter into the soil sufficiently early to complete the decomposition process well before the sowing of the crop.

Two other groups of bacteria intimately linked with the nitrogen problem of the soil take up free nitrogen from the air and convert it into nitrogenous compounds for the use of crop plants. One group (genus *Rhizobium)* functions in symbiosis with leguminous plants, and the other (belonging to genus of *Azotobacter)* fixes free nitrogen independently of the legumes. The symbiotic bacteria penetrate the rootlets and induce the plants to produce nodules at the points of entry. The bacteria grow and multiply inside the nodules, "getting their carbohydrate and mineral food from the plant and nitrogen from the air to form nitrogenous organic compounds, e.g. proteins, which then become available to the host plant. The amount of nitrogen added to the soil varies from 50-150 kg per hectare. In the absence of the nitrate-fixing bacteria, no nodules will be formed by the leguminous plants and no atmospheric nitrogen will be fixed. It has been found that the nodule organisms belong to several distinct strains, each of which is capable of producing nodules on a specific group of legumes. It has also been discovered that the presence of the most efficient bacteria for a given legume crop can be assured by inoculating the seed with a culture of the proper strain artificially grown in the laboratory.

Azotobacter and other non-symbiotic nitrogen-fixing bacteria work independently of any host crop. Under optimum laboratory conditions, *Azotobacter* has been found to fix a considerable amount of nitrogen.

Actinomycetes : They are similar in size to bacteria but resemble moulds in their growth and physiology. They can grow in the deeper layers of the soil and under drier conditions, and need less nitrogen. They are also said to synthesize chemical substances which produce the familiar odour, when the sun-baked soil of a cultivated field receives the first shower of rain.

Fungi : These organisms produce microscopic threads called mycelia, which may be found in the disintegrating organic matter on the surface of the soil or on plant roots in the upper strata below the surface. Many of the fungi are harmless saprophytes, living on dead organic matter; others are parasites which attack live plants and produce highly destructive epidemic diseases. Still others are parasites that can also live in the absence of host plants.

Algae : They are microscopic or larger plants containing chlorophyll. They are found in large numbers in the top layer of constantly moist soils, such as paddy fields. Some of them have been shown to be capable of fixing atmospheric nitrogen.

Protozoa : Soil protozoa are unicellular animals, feeding either on soil organic matter or on bacteria, thus regulating the number of the latter in the soil. The larger forms are found in very wet conditions or in swamps.

Worms and Insects : Most of the soils are inhabited by worms, insects and other animals of different sizes. They feed on plant remains, which are ingested along with large quantities of earth. Their excrements in the form of oasts is brought up and deposited on the surface.

Insects and some other larger animals promote aeration and water percolation. They make channels and burrows in the soil primarily for their protection or in quest of food.

Soil and Mapping Survey

Modern soil surveys are basically aimed at furnishing comprehensive information about soils and an inventory of soil resources of the area. They consist in studying and recording important characteristics of the soils in the field and in the laboratory, classifying them into well-defined units and locating their extent and boundaries on a map. The broad general objectives of soil surveys are both of fundamental and applied nature. Fundamentally, soil surveys help to expand our knowledge and understanding of soils as regards their genesis, development, classification and nomenclature. The applied part of the surveys includes the interpretation of soil data for use in agriculture, forestry, engineering, urban development, etc.

It gives information on how to evolve a rational land-use plan in a given area. Further, it helps to correlate the characteristics of soils of known behaviour and predict their adaptability to various uses and also their behaviour and productivity under a defined sets of management practices. It thus forms the very basis for planned land use.

Different Types of Surveys : Two main types of surveys are: (a) reconnaissance surveys, and (b) detailed surveys, depending upon the method, the intensity of the survey, the scale of the base map, and the resulting details and precision of mapping.

1. Reconnaissance Soil Survey: In the case of a reconnaissance survey, the scale used for mapping is 1:63,360 (1" =1 mile) survey of India toposheet, or 1: 50,000 or a similar-scale air photographs, wherever they are available. Mapping is done after an initial interpretation of the base maps and by constructing the initial legend. Soil series or an association of soil series is normally the unit of mapping. The intensity of profile examination or the auger-bore observation depends on the heterogeneity of the land and soil features. In a highly heterogeneous area, the number of observations

for mapping would be large and vice versa. Reconnaissance surveys and maps serve important purposes of identifying bench-mark soils, delineating problem and potential areas and providing information useful for a broad land-use planning and agricultural development at the tehsil level and upwards. Progressive abstraction, synthesis and the compilation of such tehsil soil maps lead to the generalized district, state and country -level soil maps, showing the delineation and classification of the soils at progressively higher categories.

2. Detailed Soil Survey : In the case of this type of survey, maps with a scale large enough to delineate the phases of soil types, slopes, erosion, etc. within a series are used. Cadastral maps on the scale of about 1:8000 or larger, or aerial photographs of 1:20,000 or larger are used as base maps. Depending upon the land and soil heterogeneity and the extent of details required, the profile and auger-bore observations are made at closer intervals and the mapping units at the phase level are delineated.

The detailed soil map furnishes information needed for understanding the soil and land problems and for working out measures needed for the conservation of soil, for water-use management, drainage, etc. and the maintenance of soil fertility without deterioration.

Defining Soil Survey

Classification based on Land Capabilities : The land-capability classification is an interpretative grouping of soils primarily meant for agricultural purposes and is mainly based on (i) the inherent soil characteristics, (ii) the external land features, and (iii) environmental factors. The classification of soil units mapped into capability groupings enables one to understand (i) the hazards of the soil to various factors which cause soil damage, deterioration in fertility, and (ii) its potentiality for production. The capability classification provides three major categories: (i) a capability-unit, (ii) a capability sub-class, and (iii) capability class.

1. Unit of Capability : The capability unit is a grouping of soils that have almost the same influence on production and responses to the systems of management of common cultivated crops.
2. Important Sub-class : The capability subclass is a grouping of capability units having similar limitations and hazards. Pour kinds of limitations recognized are: (i) erosion, (ii) wetness, (iii) shallowness of the root-zone, and (iv) climate.
3. Important Class : The capability class is the broadest category of classification. Eight classes are recognized and the risk of soil damage or limitations in use progressively increases from class 1 to class VIII. Soils in classes I to IV are arable. They are capable of producing commonly cultivated crops of the region under good management. Soils in classes V to VII are suited to adapted native plants, pasture or forestry. Soils of class VIII are neither suitable for agriculture nor for silviculture. The capability classes are briefly described below:

Drainage of Land

1. Class I : The soils in class I have only a few limitations that restrict their use. The soils are nearly level, deep, well drained, with good water-holding capacity. They are productive and suitable for intensive cropping.
2. Class II. : The soils in class II have some limitations that reduce the choice of crops and require moderate con servation practices to prevent deterioration, when culti-

vated. The soils may be used for raising many of the cultivated crops.

3. Class III : The soils in class III have severe limitations that reduce the choice of crops, and require special conservation practices when used for raising cultivated crops. The limitations of soils in class III are more than those in class II and, therefore, need careful management.

4. Class IV : The soils in class IV have very severe limitations that restrict the choice of crops to only a few and require very careful management. The cultivation of crops may be restricted to once in 3 or 4 years. Conservation measures are more difficult to apply and maintain in these soils.

Drainage not on Land

1. Class V : The soils in class V have little or no erosion hazard, but have other severe limitations, such as witness or overflow, as in the case of bottom lands which prevent the normal tillage for common cultivated crops. These restrictions limit their use to pasture or adaptable tree species.

2. Class VI : The soils in class VI have severe limitations to depth, slope, erosion hazard, etc. that make them unsuitable for the cultivation of normal crops. Their use is restricted to pasture and silviculture.

3. Class VII : The soils of class VII have more limitations than those of class VI and make them totally unsuitable for the cultivation of crops and restrict their use only to grazing.

4. Class VIII : The soils of class VIII have very severe limitations that preclude them from being used for agriculture or silviculture. Their use is restricted to recreation or wildlife.

Other types of specific interpretative groupings can also be developed from the data collected during the survey and can be supplemented with laboratory characterization. Thus the soil-survey data and soil maps can be interpreted for land capability, soil and land irrigability classifications, engineering Uses, etc.

Soil Categorization

Soils widely vary in their characteristics and properties. In order to establish the interrelationship between their characteristics, they require to be classified. Understanding the properties of soils is important in respect of the optimum use they can be put to and the best management requirements for their efficient and productive use. Classification helps to reduce the study of the number of individuals to a few well-defined units. It helps to develop a legend for mapping soils during the surveys. It helps to group together such soils as have comparable characteristics so that the existing knowledge regarding them is presented in a systematic manner.

To classify soils and group them together in a meaningful manner, different systems of soil classification have been used from time to time. These systems have varied over a period of time, having been drawn up to meet the requirements and the immediate purposes of their use. As a knowledge of the soils which helps one to understand their genesis has developed the systems of classification have also been developed, keeping pace with the requirements.

The modern system of classification, "Soil Taxonomy" developed by the USDA has been recommended for adoption all over the world and in this country as a result of the decision taken at the All-India Workshop held in 1960. This is a multi-category system which has six categories, namely, order, sub-order, great group, sub-group, family and series. The system is hierarchical in that an order is divided into sub-orders, and sub-orders into great groups, great groups into sub-groups, etc. up to the series level. Thus the number in the higher categories is fixed, whereas in the lower categories, it is variable. The soil series is the basic unit of classification and in all our surveys, the soil series are first identified and described.

Whereas phases of soil series are used as mapping units in detailed soil surveys, the associations of soil series are used as mapping units in reconnaissance soil surveys. The progressive abstraction and compilation of reconnaissance maps help one to construct small-scale maps at the state and country levels with the associations of higher categories as the mapping units. Thus, for example, great groups or the association of great groups are the units delineated at the state or country level.

Brief descriptions of the major soil groups, along with their modern nomenclature, are given below :

Depositional Soils : The alluvial soils include the deltaic alluvium, calcareous alluvial soils, coastal alluvium, and coastal sands. This is by far the largest and most important soil group of India contributing the largest share to its agricultural wealth.

In this immense tract, though a great deal of variation exists, the main features of the soils are derived from the deposition laid by the numerous tributaries of the Indus, the Ganges and the Brahmaputra systems. These streams, draining the Himalayas, bring with them the products of weathering of rocks constituting the mountains in various degrees of fineness and deposit them as they traverse the plains.

Geologically, the alluvium is divided into *khadar* i.e. the newer alluvium of sandy, generally light-coloured and less *cankary* composition, and *bhangar*, i.e. the older alluvium of more clayey composition, generally dark and full of *kankar*. The soils differ from the drift sand to loams and from fine silts to stiff clays. A few occasional pebblebeds are also present. The presence of impervious clays, in part, obstructs the drainage and to some extent promotes the accumulation of injurious salts of sodium and magnesium and these make the soil sterile.

The formation of hard-pans, at certain levels, in the soil profile through the binding of soil grains by the infiltrating silica or calcareous matter, forming an impervious layer, is often observed in these alluvial soils. Layers of *kankar* in the Indo-Gangetic alluvium of Uttar Pradesh and West Bengal and also occasionally layers composed of impure iron oxides are instances of the formation of hard-pans.

The most important characteristic of the soil of Assam is its acidity. Generally, those on the old alluvium and hills are more acidic than the new alluvial soils along the river banks. The latter are often neutral or alkaline. The soils of the Brahmaputra Valley are sandy; their available and total potash contents are fairly good; their P_2O_5 content is good and the percentages of organic matter and nitrogen in them are fairly moderate. The soils of the Surma Valley are fine in texture.

In West Bengal, the portions of Murshidabad, Bankura, the whole of Burdwan, and the western half of Midnapore, comprising the tract known as the Rarh region, are composed mainly of the old alluvium. There is hardly any regularity in the manner of deposition of the river-borne materials. Some of the deposits, which had been laid down very early, have naturally been subjected to the climatic and other influences, leading to soils which may be different from one another in texture, colour profile, chemical composition and mechanical and other physical properties.

The alluvial soils of Bihar may be divided into two main divisions, depending on the distinctness of their characters, e.g. (a) the alluvium north of the Ganges; and (b) the alluvium south of the Ganges.

(a) The northern alluvium comprises the area between the Himalayas in the north and the Ganges in the south.

The soil is alluvial, with a calcareous belt in the form of a triangle in the west, and broken inundated areas in the middle; these areas remain flooded for different periods in the year. The soils are sandy loam to clayey loam and neutral to alkaline. Their CaO ranges from 0.5 to 20 per cent. They are rich in total and available potash, but are deficient in phosphorus.

(b) The southern alluvium comprises the area between the Ganges in the north and the hilly region in the south. The soils vary in colour and texture from light-greyish loams to heavy black clays. The middle of the area has a depressed feature, which gets flooded during the monsoon and looks like a big lake. The area has been sub-divided into: (1) highland soils, (2) soils liable to inundation, (3) saline soils, and (4) *diara* land soils

The soils of Uttar Pradesh are divided into four classes: (a) the alluvium of the west and north-west, lighter in texture, (b) the alluvium in the centre, with texture intermediate between light and heavy, and (c) the alluvium in the north-east developed on a calcareous parents material.

The soils contain varying amounts of $CaCO_3$ and soluble salts and are neutral to alkaline. The lime content usually increases at lower depths. They are generally poor in P_2O_5, nitrogen and organic matter.

On the coast of Orissa, there are stretches of sand and sand-hills, alternating with deltaic swamps. Behind this coastal belt is an area of cultivated alluvial and lateritic formations. Soils are both sandy and of finer texture; there is sufficient potash but not enough P_2O_5.

The alluvial soils of Tamil Nadu are found in the deltaic areas and along the coast. A section of its profile reveals a deposition of alternate layers of sand and silt, as they are brought in by the rivers. The composition of the strata varies with the nature of the silt brought by the rivers, and in turn, varies with the catchment areas and the tracts through which they flow. The stratified deposition is confined to areas very near the river courses. But away from the rivers, the soils are heavy throughout, the texture ranging from clay loam to heavy clay through silty clay. In such cases, sandy layers occur at very low depths.

In the Gujarat state, the alluvial soils are confined to the northern Gujarat tract, Ahmedabad and Kaira districts and they are locally known as *goradu*. The *gorat* soil of Baroda corresponds to the older alluvium, consisting of brown clay with *kankar*. Those from the recent depositions are known as *bhata*. The soils, to a great extent, are of secondary deposition, fairly deep, poor in organic matter and nitrogen, but fairly rich in P_2O_5 and K_2O.

The light-sandy, red and yellow soils found in the Mahanadi basin (Madhya Pradesh), including the Balaghat and three districts of Durg, Raipur and Bilaspur, are of eluvial origin.

The soils of the Punjab and Haryana plains belong to the same class of alluvial soil that is typical of the Indo-Gangetic plains. The majority of the soils are loams or sandy loams consisting of a soil crust of varying depth. Soluble salts are present in considerable amounts. The lower layer contains *kankar* nodules. Owing to the presence of sodium in the clay complex, the soils are generally alkaline. They are adequately supplied with phosphorus and potash, but are deficient in organic matter and nitrogen.

In Kerala, there are two types of alluvial soils on river banks, viz. the coastal alluvium and the alluvium. In central Kerala, the width of the coastal eluvial tracts increases, whereas in the north and south, they are comparatively narrow. The alluvial soils of Kuttanad form a low lying area, believed to be once a part of the sea and later filled up by the silt carried down by the Pampa and other rivers. The coastal alluviums are sandy, having a low water-holding capacity and a low nutrient status. The alluviums on the banks of the rivers are fertile.

Regur Soils : These soils vary in depth from shallow to deep. The typical soil derived from the Deccan trap is the *regur* or black cotton soil. It is common in Maharashtra, western parts of Madhya Pradesh, parts of Andhra Pradesh, parts of Gujarat and some parts of Tamil Nadu. It is comparable with the 'chernozems' of Russia and the 'prairie soil' of the cotton-growing states of the United States, especially the 'black adobe' of California. It is derived from two types of rocks, the Deccan and the Rajmahal trap and ferruginous gneisses and schists occurring in the Tamil Nadu state under semi-arid conditions. The former attains sometimes considerable depths, whereas the latter are generally shallow. Generally, there is no change in colour up to a thickness of two to three metres.

Many black-soil areas have a high degree of fertility, but some, especially in the uplands, are rather poor. They are somewhat sandy on the slopes and uplands and are moderately productive. In the broken country, between the hills and the plains, they are darker, deeper and richer and are constantly enriched by the additions washed down from the hills.

Black soils are highly argillaceous, very fine-grained and dark and contain a high proportion of calcium and magnesium carbonates. They are very tenacious of moisture and exceedingly sticky, when wet. Owing to considerable contraction on drying, large and deep cracks are formed. These soils contain abundant iron and fairly high quantities of lime, magnesia and alumina. Potash has a wide range. They are poor in phosphorus, nitrogen and organic matter. In all *regur* areas, in general, and in those derived from ferro-magnesium schists, in particular, there is generally a layer rich in *kankar* nodules formed by the segregation of calcium carbonate at some depth below the surface and above the weathered rock. The soils are generally rich in montmorillonitic and beidellitic group of clay minerals.

In Maharashtra, the soils derived from the Deccan trap occupy quite a large area. On the uplands and slopes, soils are light-coloured, thin and poor. On the lowlands and in the valleys, deep and relatively clayey black soils are found. Along the Ghats, the soils are very coarse and gravelly. In the valleys of the Tapti, the Narmada, the Godavari and the Krishna rivers, heavy black soil is often 6 metres deep. The subsoil contains a good amount of lime. Outside the Deccan trap area, the black cotton soil predominates in Surat and Broach districts. Degraded solonized black soils, locally known as *chopan,* occur in areas in the canal zones of the Deccan in Maharashtra.

In Tamil Nadu, the black soils are either deep or shallow and may or may not contain gypsum in their profiles. Soils are fine-textured, have high pH (8.5-9) and are rich in lime (5-7 per cent). They have low permeability and high values of hygroscopic co-efficient, pore space, maximum water-holding capacity and true specific gravity. Black soils have generally a high base status and a high cation-exchange capacity of 40 to 60 m.e. per 100 g. The analysis of clay fractions shows that the iron content varies from 10 to 13 per cent and the SiO_2 / R_2O_2 ratio varies from 3 to 3.5, CaO and MgO contents are high. The soils are found to be formed from a variety of rocks which include traps, granites and gneisses.

In Madhya Pradesh, two distinct kinds of black soils are found, viz. (i) deep heavy black soils covering the Narmada Valley, and (ii) shallow black soils in other areas. The cotton-growing areas are mainly covered by the deep heavy black soils, but there are also soils of lighter texture. The clay content usually varies from 35 to 50 per cent and the SiO_2/R_2O_3 ratio from 3 to 3.5. The organic matter content is low.

The black soils of Karnataka are fine-textured with varying salt concentration. The soils are generally rich in lime and magnesia, the SiO_2/R_2O_3 ratio of clay fraction is 3.6.

Red Type Soils : The soils comprise vast areas of Tamil

Nadu, Karnataka, Goa, Daman and Diu, south-eastern Maharashtra and eastern Andhra Pradesh, Madhya Pradesh, Orissa and Chhotanagpur. In the north, the red-soil area extends into and includes the greater part of Santhal Paraganas in Bihar, the Birbhum district of West Bengal, the Mirzapur, Jhansi and Hamirpur districts of Uttar Pradesh.

The ancient crystalline and metamorphic rocks on meteoric weathering have given rise to the red soils. The colour of the soil is due to the wide diffusion of iron rather than to a high proportion of it. The soils grade from poor thin gravelly and light-coloured varieties of the plains and valleys. They are generally poor in nitrogen, phosphorus and humus. These soils are poorer in lime, potash, iron oxide and phosphorus than the *regur* soils. Many of the so-called red soils of southern India are not red. On the other hand, some red soils are of lateritic origin and of a quite different nature.

The clay fraction of the red soils is rich in kaolinitic type of mineral. Red soils have also been found under forest vegetation. Red and yellow soils are also seen side by side. Very little is known about the yellow soils. Their colour is probably due to a higher degree of hydration of the ferric oxide in them than that in the red soils.

Morphologically, the red soils can be divided into two broad sub-groups : (i) red loams, characterized by argillaceous soils with a cloddy structure and the presence of only a little concretionary material; and (ii) red earths where the top soil is loose and friable and rich in secondary concretions.

The red soils in Tamil Nadu occupy the largest area and constitute nearly two-thirds of the cultivated area. They are all *in situ* formations from the underlying rock under the influence of climatic conditions. The rocks are micaceous or red granites; the latter are acidic. The soils are rather shallow, open in texture, have a pH ranging from 6.6 to 8, they have a low base status and their exchange capacity is low. They are also deficient in organic matter and poor in plant nutrients. An analysis of their clay fractions reveals SiO_2/R_2O_3 ratios of 2.5 to 3.0.

The predominant soil in the eastern tract of Karnataka is the red soil overlying the granite from which it is derived. Especially in the districts of Bangalore, Kolar, Mysore, Tumkur and Mandya, this is the chief type varying in depth. There are shades of red and these shades pass on to the yellows. Loamy-red soils are predominant in the plantation districts of Shimoga, Hassan and Kadur. They are rich in the total and available K_2O and contain sufficient amounts of the total P_2O_5 (0.05-0.3 per cent) and their lime content varies from 0.1 to 0.8 per cent. Nitrogen is below 0.1 per cent. Iron and alumina are high, being 30-40 per cent.

A broad strip of the area running between the eastern and western parts of Coorg is red loam, easily drained, and with a fairly dense growth of trees.

The acid soils towards the south of Bihar, viz. those of Ranchi, Hazaribagh, Santhal Paraganas, Manbhum and Singhbhum, are red soils. The pH of the soils varies from 5 to 6.8. Another distinctive feature is the higher percentage of acid-soluble Fe_2O_3 than that of Al_2O_3. Available potash is quite sufficient, but P_2O_5 is low.

In West Bengal, the red soils (sometimes misrepresented as laterites) are the transported soils from the hills of the Chhotanagpur plateau.

A typical profile of red soil at the Chandkhuri Farm, Raipur, Madhya Pradesh, reveals that the percentage of concretions increases down the profile. The total exchangeable cations are about 20 m.e./100 g. The SiO_2/R_2O_3 ratio of the clay fraction varies between two and three and the C/N ratio is approximately ten.

A part of the Jhansi district in Uttar Pradesh comprises rod soils. There are two types, locally known as *parwa* and *rakkar*.

The *parwa* is a brownish-grey soil varying from good loam to sandy or clay loam. The *rakkar* is the true red soil which is generally not useful for cultivation. The soils of Banaras and Mirzapur developed on the Vindhyan parent materials, have also been classified as tropical and subtropical red loams.

In the Telangana Division of Andhra Pradesh, where the predominating geological formation is granite and a gneissic complex, both red and black soils predominate. The red soils or *chalkas* are sandy loams located at higher levels. Such soils are utilized for the cultivation of *kharif* crops.

Laterites and Lateritic Soils : Laterite is a formation peculiar to India and some other tropical countries, with an intermittently moist climate. It is a compact to vesicular rock composed essentially of a mixture of the hydrated oxides of aluminium and iron with small amounts of manganese oxides, titania, etc. It is derived from the atmospheric weathering of several types of rocks. Under the monsoon conditions of alternating wet and dry seasons, the siliceous matter of the rooks is leached away almost completely during weathering.

The laterite may become broken off and be carried to lower levels by the action of streams and, when redeposited at lower levels, may become cemented again into a compact mass by the segregative action of the hydrated oxides, including sand grains of quartz and other minerals. Thus there are high-level laterites resting on the rocks at whose expense they have been formed and the low-level laterites formed in the usual way of detrital deposits.

Laterites are specially well developed on the summits of the hills of Karnataka, Kerala, Madhya Pradesh, the Eastern Ghat regions of Orissa, Maharashtra, West Bengal, Tamil Nadu and Assam. All lateritic soils are very poor in lime and magnesia and are deficient in nitrogen. Occasionally, P_2O_5 may be high, probably being present in the form of iron phosphate, but K_2O is deficient. There is occasionally a higher content of humus.

In Tamil Nadu, there are both high-level and low-level laterites which are formed from a variety of rock materials under peculiar climatic and weather conditions. They are both *in situ* and sedimentary formations and are found all along the West Coast, where rainfall is heavy and humid climate prevails, and also in some parts of the East Coasts.

On the laterites, at lower elevations paddy is grown, whereas on those situated at higher elevations, tea, cinchona, rubber and coffee are grown. The soils are rich in nutrients and contain 10-20 per cent organic matter. The pH is generally low, particularly of the soils under tea (pH 3.5-4), and the higher the elevation, the more acidic the soils are.

In the laterite soils of Ratnagiri (Maharashtra), the coarse material is found in large quantities. These soils are rich in plant food constituents, except lime.

In Kerala, both high-level and low-level laterites occur. The high-level laterites growing plantation crops are rich soils because of their proper management. The laterites on lower elevations have a poor nutrient status. Those of the West Coast generally grow plantation crops, e.g. tea, rubber, cinchona, coconut and arecanut, but at low elevations, paddy is also grown. The soils are generally poor in N P K and organic matter, the pH ranging from 4.5 to 6.0. SiO_2/R_2O_3 ratio gradually increases with the decreasing elevation (from 1.32 to 2.08).

The laterite soils in Karnataka occur in the western parts in the districts of North Kanara and South Kanara, Shimoga, Hassan, Kadur and Mysore. All the soils are comparable with the laterites and similar formations in Malabar, and Nilgiris District. These soils are very low in bases, because of the severe leaching and erosion., These soils are also poor in P_2O_5. Their pH is not so low as that of plantation soils.

In West Bengal, the area between the Damodar and the Bhagirathi is interspersed with some basaltic and granitic hills, with laterite capping. In these laterite soils, the SiO_2/Al_2O_3 ratio of the clay fraction is quite high. The percentages of K_2O, P_2O_5 and N are low, showing considerable leaching and washing out of these substances as a result of chemical weathering. In Bihar, the laterite occurs principally as a cap on the higher plateaus, but is also found in fair thickness in some valleys.

The laterites of Orissa are found largely capping the hills and plateaus occasionally in considerable thickness. Large areas in Khurda are occupied by laterites ; those of Balasore are gravelly and appear to be detrital.

Desert Soils : A large part of the arid region, belonging to western Rajasthan, Haryana, Punjab, lying between the Indus river and the Aravalli range is affected by desert conditions of geologically recent origin. This part is covered under a mantle of blown sand which, combined with the arid climate, results in poor soil development. The most predominant component of the desert sand is quartz in well-rounded grains, but feldspar and hornblende grains also occur with a fair proportion of calcareous grains.

The desert proper, owing to the physiographic conditions of its situation, though lying in the track of the south-westerly monsoon, receives little rain. The sands which cover the area are partly derived from the disintegration of the subjacent rocks, but are largely blown in from the coastal regions and the Indus Valley. Some of these soils contain high percentages of soluble salts, possess high pH, have a low loss on ignition, a varying percentage of calcium carbonate and are poor in organic matter.

The Rajasthan desert is a vast sandy plain, including isolated hills or rock outcrops at places. Though, on the whole the tract is sandy, the soil improves in fertility from west and north-west to east and north-east. In many parts, the soils are saline or alkaline, with unfavourable physical conditions and high pH.

The classification of the soils, according to soil taxonomy, along with traditional nomenclature, is given below:

Major group	Soil taxonomy orders
Alluvial soil	Entisol
	Inceptisol
	Alfisol
Black soil	Vertisol
	Inceptisol
	Entisol
Red soil	Alfisol
	Inceptisol
	Ultisol
Laterite soil	Alfisol
	Ultisol
	Oxisol
Desert soil	Entisol
	Aridisol

Problem Soils : The problem soils are those which, owing to land or soil characteristics, cannot be economically used for the cultivation of crops without adopting proper reclamation measures. Highly eroded soils (sheet and gully), ravine lands, soils on steeply sloping lands, etc. constitute one set of problem soils. The shallow soil depth, deep gullies, steep and complex slopes are some of the problems which require to be tackled in such areas. Their reclamation may involve massive earth-moving operations, terracing, afforestation or plantation to maintain permanent cover with grasses, depending upon the intensity of the problem and the nature of the terrain and soil conditions. The potentiality of the land, present land use, cost of operations and other socio-economic factors of the region are some of the factors which have to be taken into consideration.

Acid saline and alkali soils constitute another set of problem soils, in the case of which acidity, soluble salts and exchangeable sodium limit the scope of cultivation.

Soil having Acidity : Although soils having pH below 7 are considered to be acidic from the practical standpoint, those with pH less than 5.5 and which respond to liming may be considered to qualify to be designated as acid soils. In the classification of soils both the percentage base saturation and the pH are used as criteria to distinguish acid soils from non-acid ones.

Acid soils occur widely in the Himalayan region, the great eastern plains of extra-peninsular India, the peripheral peninsula and the coastal plains, including the Gangetic delta. They are found to occur on different geological formations under varying physio graphical, climatic and vegetational environment. In all these regions, however, the rainfall component of the climate appears to have a dominating influence on the formation of acid soils.

In humid regions, where rainfall is high, the soluble bases formed in the course of weathering of rocks are leached down and carried away by the drainage waters. The continued leaching of soils results in the replacement of calcium, magnesium, potassium and sodium ions by hydrogen ions and the formation of acid soils with low pH. In acid soils, the dissolution of alumino silicate minerals occur and the aluminium ions, thus released, increase the acidity owing to hydrolysis. Similarly, humus and hydrous oxides contribute to soil acidity at low pH.

Soil acidity exceeding a particular limit is injurious to plant growth. The availability of certain nutrients, particularly phosphorus, calcium and magnesium, becomes low with increasing acidity. On the other hand, in acid soils, ions, e.g. those of aluminium, iron, manganese and copper, may be found in the dissolved form in quantities sufficient to become toxic.

Similarly, most of the desirable soil microbiological processes, such as the beneficial activities of *Azotobacter* and nodule-forming bacteria of legumes, are adversely affected as the acidity increases. The satisfactory granulation of soils also becomes difficult to achieve. It is, therefore, necessary to correct soil acidity for cultivating such soils profitably.

On the basis of pH measurements, the degree of soil acidity may be approximately indicated as below :

pH	*Reaction*
6.6 to 7.5	Nearly neutral
6.1 to 6.5	Slightly acid
5.6 to 6.0	Medium acid
5.1 to 5.5	Strongly acid
4.6 to 5.0	Very strongly acid
4.5 and below	Extremely acid

It should be stressed here that the pH value of the soil indicates only the active acidity. For remedial measures, the total acidity should be considered, which is briefly discussed below :

Lime Necessity : Acidity in the soil systems can be conveniently classified as active or potential acidity. Active acidity includes hydrogen ions in the solution phase and is determined by pH measurements. The potential acidity may be considered to be the exchange acidity and makes up the bulk of the total acidity which is many times greater than active acidity. In the soil, an equilibrium exists between the active acidity and the potential acidity.

The lime requirement of a soil necessary to neutralize the total acidity may be defined as the amount of liming material that must be added to raise the pH to some prescribed value. This value is usually in the range of pH 6 to 7, since this range is easily attainable within the optimum growth range of most crops. The pH measurements of the soil are widely used for estimating the lime requirement. The basis for this method is that in acid soils, a relationship exists between the pH and the percentage base saturation of the soil. Once this relationship is known, it is useful for estimating the lime requirement in the laboratory. However, with this "lime requirement' under field conditions, the predicted pH is not generally attained. Hence a "liming factor" of 1.5 to 2 is commonly used to achieve the desired results, i.e. the lime requirement, as determined in the laboratory, is multiplied by a factor of 1.5 to 2. The approximate amount of limestone required to raise the pH to the neutral level for some soils is given in Table.

The Quantity of finely Ground Limestone Required to Raise the pH Value of the Ploughed Layer (about 15-18 cm, of the Soil from the Value indicated to a pH of 6.5

	Quantity of limestone required (tonnes/ha)		
Soil	*From pH 4.0*	*From pH 4.5*	*From pH 5.5*
Warm humid plains			
Sands and loamy sands	3.76	2.5	1.25
Sandy loams	–	5.0	2.5
Loams and silt loams		8.75	5.0
Clay loams	—	12.5	7.5
Cool temperate hilly region			
Sands and loamy sands	7.5	5.0	2.5
Sandy loams	–	7.5	5.0
Loams and silt loams		11.25	7.5
Clay loams	—	15.0	8.75
Valley soils			
Organic, water-logged	22.6	17.5	11.5

The liming material should be evenly broadcast and worked into the soil several weeks before sowing a crop to allow time for completing the reaction. Keeping the land moist hastens the exchange process. Although liming once in 5 years may serve the purpose, the frequency of liming should be determined by making periodic pH measurement.

Table gives only a general picture of the lime requirements on the basis of pH and texture. In several states of India, lime requirements have been worked out for specific soils. Hence, these results should be checked by making a reference to the department of agriculture or the agricultural university of a state.

Crops and Acid Tolerance : Many of the principal crops and vegetables are sensitive to acid soils and suffer injury when grown on them. The optimum pH ranges of some of the crops is given in the Table. This information should be helpful in estimating the lime requirements for a particular crop.

Soil Reaction Preferences of Crop Plants

Crops	*Optimum pH range*
Cereals	
Maize (*Zea mays*)	6.0-7.5
Millets	5.0-6.5
Rice (*Oryza sativa*)	4.0-6.0
Sorghum	6.0-7.5
Wheat	6.0-7.5
Barley	6.0-7.5
Oats	5.0-7.7
Legumes, field and garden crops	
Field bean (*Phaseolus vulgaris*)	5.5-7.0
Berseem (Trifolium alexandrinum)	6.0-7.5
Groundnut (*Arachis hypogaea*)	5.3-6.6

Contd...

Crops	*Optimum pH range*
Soyabean (*Glycine max*)	5.5-7.0
Pea (*Pisum sativum*)	5.5-7.0
Lentil, gram, etc.	5.5-7.0
Miscellaneous field crops	
Sugarcane (*Saccharum officinarum*)	6.0-7.5
Cotton, upland (*Gossypium hirsutum*)	5.0-6.5
Potato (*Solanum tuberosum*)	5.0-5.5
Tea (*Thea sinensis*)	4.0-6.0

Saline and alkali or sodic soils. In many arid and semi-arid areas of India, crop production is limited because of salinity or alkalinity or both. It is estimated that about 7 million hectares in the country have either gone out of cultivation or this area produces low yields of crops. The area in different states is given in Table.

The Extent of Saline and Alkali Soils in India

State	*Area under saline/ alkali soils (million ha)*
Uttar Pradesh	1.280
Gujarat	1.200
West Bengal	0.840
Rajasthan	0.720
Punjab	0.680
Maharashtra	0.528
Haryana	0.520
Karnataka	0.400
Orissa	0.400
Madhya Pradesh	0.240

Contd...

State	*Area under saline/ alkali soils (million ha)*
Andhra Pradesh	0.024
Delhi	0.016
Kerala	0.016
Bihar	0.004
Tamil Nadu	6.872
	(or about 7 million hectares)

Classes of saline and alkali soils. Three classes of saline and alkali soils are recognized. They are briefly described below :

1. Saline Soils : The soils containing toxic concentrations of soluble salts in the root-zone are called saline soils. Electrical conductivity in the saturation extract of such soils taken as a measure of salts is greater than 4.0 mmhos/cm. Exchangeable sodium percentage is less than 15 and the pH is less than 8.5. The soluble salts mainly consist of chlorides and sulphates of sodium, calcium and magnesium. Because of the white encrustation due to salts, the saline soil is also called white alkali.

2. Non-Saline Soil : These soils do not contain any large amount of neutral salts and, as such, the electrical conductivity is less than 4 mmhos/cm. The detrimental effect of alkali soil on plants is largely due to toxicity of a high amount of exchangeable sodium and the pH. Alkali soils have an exchangeable sodium percentage of more than 15 and a pH greater than 8.5. Such soils have low infiltration rate and the physical condition is unfavourable. Because of high alkalinity, resulting from sodium carbonate, the surface soil is discoloured and black, and, hence the term black alkali is frequently used to designate the non-saline alkali soil.

3. Saline-alkali soils : This group of soils is both saline and alkali. They have appreciable amounts of soluble salts, as indicated by the electrical conductivity values of more than 4 mmhos/cm. Also, the exchangeable sodium percentage is greater than 15. The pH, however, is likely to be less than 8.5.

The soil salinity or alkalinity or both have many adverse effects, which are summarized below :

1. Causing low yields of crops or crop failure in extreme cases.
2. The limiting of the choice of crops, because some crops are sensitive to salinity or alkalinity or to both.
3. Rendering the quality of fodder poor as, at times, the fodder grown on alkali soils may contain a high amount of molybdenum and a low amount of zinc, causing nutritional imbalance and diseases among live-stock.
4. Creating difficulties in the construction of buildings and roads and their maintenance.
5. Causing excessive runoff and floods owing to low infiltration, resulting in damage to crops in the adjoining areas.

Salinity and its Causes : In arid and semi-arid areas, salts formed during the weathering of soil minerals are not fully leached. During the periods of higher-than-average rainfall, the soluble salts are leached from the more permeable high-lying areas to the low-lying areas, where, if the drainage is restricted, salts accumulate on the surface as water evaporates. The excessive irrigation of the uplands containing salts thus results in the accumulation of salts in the valleys. In areas having a salt layer at lower depths in the profile, seasonal irrigation may favour the upward movement of the salts. Salinization is also caused owing to the irrigation of soils with saline water. In all these cases, restricted drainage is usually the main reason. Rise in the water-table within 2 metres of the surface due to irrigation, the obstruction of natural drainage because of developmental activities, e.g. roads and canals, and the siltation of natural drainage may also cause soil salinity. In the coastal areas, the ingress of sea-water induces salinity in the soil. When sodium ions predominate in the soil solution, and carbonates are present, alkali soils are formed.

Reclamation : Different types of saline and alkali soils may occur singly or in association, depending upon their mode of formation and physiographic position. Also, the degree of salinity or alkalinity may vary. Thus the salt affected soils differ a great deal in their physico-chemical characteristics and, as such, methods of their reclamation also differ.

If the problem is only of salinity, the salts need to be leached below the root-zone and not allowed to come up. In practice, however, this might be difficult to accomplish, especially in deep and fine-textured soils containing more salts in the lower layers. Under these conditions, a provision of some kind of subsurface drains becomes important. If the soil contains a sandy layer at a lower depth, the leaching of the salts below this layer will check the rise of salts.

The reclamation of alkali soils needs the addition of a soil amendment, containing soluble calcium salts. The commonly used amendment is gypsum. In the course of reclamation, sodium on the exchange complex is replaced by calcium. The sodium salts, thus formed, are leached down.

The number and frequency of leaching, the quantity of gypsum to be added and the techniques involved vary from region to region, depending upon the clay mineralogy of the soils, the intensity of the problem, the subsequent use of the soils, the availability and quality of irrigation water and the economics of these operations. Hence, the state authorities engaged in this work should be consulted to draw a schedule of operations for reclamation. The operations may also include other simple agronomic techniques to reclaim soils and to know the crops and varieties that may be grown and other management practices that may be followed on such soils.

Soil Management

Successful farming does not merely depend on the knowledge of physical, chemical and biological properties of the soil. It is a matter concerning both soil and soil management The most important consideration in soil management is the correct application of the relationship between the soil and the crops to be grown. Although the problems of soil management vary according to the soils and their situations, the climatic conditions and the crops to be growth yet there are fundamental factors which govern the choice of a soil-management practices.

Good soil tilth is the first feature of good soil management. It means a suitable physical condition of the soil and implies, in addition, a satisfactory regulation of soil moisture and air. The maintenance of soil organic matter which encourages granulation is an important consideration of good tilth. Tillage operations and timings should be so adjusted as to cause the minimum destruction of soil aggregates. Good tilth minimizes erosion hazards.

The choice and sequence of adaptable crops or crop rotation are other very important considerations. These issues are related to climate, particularly rainfall and its distribution and the characteristics of the soil profile, including drainage and available soil moisture. A proper sequence of crops and varieties greatly influences the soil physically, chemically and biologically. It would be more realistic to evolve cropping patterns according to the land-capability classification but where such information is not available (since a greater part of the country has not yet been surveyed systematically), the recommendations of the local agricultural agencies may have to be followed. The cropping pattern and management practices should be based on the principles of soil and water conservation and efficient moisture utilization. In irrigated areas, special management practices become necessary to avoid salinity, alkalinity and water-logging. The productive capacity of the soil should be improved and maintained by providing adequate organic manures and plant nutrients through fertilizers and by including legumes in the rotation. Similarly, the provision of irrigation facility in the semi-arid and arid areas, the adoption of different remedial measures against excessive salinity and alkalinity, or acidity in humid areas, the use of specific soil amendments to correct plant-nutrient imbalances and the application of trace elements, where they are deficient, are some of the other improvement measures needed in different and special circumstances.

Economic plant-protection measures should form part of the management practices in the cropping system. This object may be achieved by regulating the cultural practices or by using recommended practices regarding the application of fungicides, insecticides, etc.

The economics of selected management practices is of vital importance. Unless the recommended practices are economically profitable, they are not likely to be adopted, howsoever sound they may be on scientific considerations. Therefore, a package of practices for the integrated land-use plan, including all the above points and yet economically profitable, is necessary. Such an integrated package of practices is now available for different areas in the country through the all-India co-ordinated schemes, operational research projects and several other schemes both at the centre and state levels.

Before proceeding to suggest changes in soil management for an individual holding or a group of holdings, all the available information on different items mentioned above should be collected from field surveys, the nearest farms of research stations, enterprising farmers and extension workers and analysed to arrive at a rational package of practices for evolving suitable cropping systems, such a package of practices should be developed to cater for the needs of all kinds of farmers—large, small and marginal.

Testing Process of Soil

It is well known that one of the chief factors responsible for obtaining satisfactory crop yields is the presence of essential plant nutrients in the soil in adequate quantities and in readily utilizable form. For maximum production and rational soil management, a knowledge of the fertility status and physical properties of a soil is essential. Soil-testing is one of the methods of determining the fertility status of the soils, so that recommendations in regard to deficient nutrients or soil amendments can be made. In fact, soil-testing forms an essential part of any scheme of agricultural development. Soil-testing laboratories have been established in almost all the states, covering all districts in them, where soils are analysed quickly and recommendations are made in respect of the fertilizer requirement for different crops.

Sampling of Soil: Soil tests and their interpretations are based on the samples analysed. It is, therefore, important that the soil samples should be properly collected and should represent the area. Soil tests and their interpretations are as reliable as the samples drawn. For routine soil-testing, the field is traversed, variations in slope, colour, texture, management and cropping pattern are noted down and it is divided into portions according to the variations. Separate samples are collected for each of the portions.

Samples are taken, using proper tools, e.g. a soil tube an auger, a spade or a pick-axe for the plough depth at several spots and then composited. The soil is mixed thoroughly and spread on a clean sheet of paper or on a piece of cloth and divided into four equal parts. Two opposite quarters are rejected and samples from the other two are mixed and the procedure is repeated till the desired size of the sample is obtained (1/2 kg) which is collected in a plastic-bag and placed in a box or cloth-bag. Moist samples are dried before collecting; care is taken to avoid contamination. The bag or the box is well labelled and sent to the nearest soil-testing laboratory, along with the information sheet (to accompany each sample), for which the proforma is given below :

Sample Information Data

Sample No. *Date of collection of the soil sample*

1. Name of the cultivator
2. Address
3. Village
   Taluk
   District
4. Type of soil
5. Identification of the field (Survey No.)
6. Proposed crop and variety
7. History of the field for the past 3 years :

Month and year of planting	*Month and year of harvesting*	*Crop and variety*	*Yield*	*Kind of fertilizer used*	*Quantity*

8. Wet/dry/garden
9. Extent of the sample areas
10. Depth of sampling
11. Number of samples taken to make a composite sample.
12. Check the following :

    Was the stand of the previous crop uniform ? Yes or No.

Texture of the top-soil: Sandy, loam, clay loam

Slope : Level, steep, moderate

Drainage: Satisfactory, moderate, water-logged

13. Any other useful information
14. Quantity of farmyard manure available

In collecting the soil samples, the help of the local officials, namely, the village level worker or the extension worker can be taken. He can also help to collect the data required to fill the information sheet.

Test of Soil : In the soil-testing laboratory, soil samples are analysed by using standardized rapid methods for the following items :

(i) pH or the soil reaction which indicates whether the soil is acidic, alkaline or normal.

(ii) Total soluble salts, as determined by electrical conductivity which indicates the degree of salinity, alkalinity, etc. of the soil.

(iii) Organic carbon which is a measure of available nitrogen.

(iv) Available phosphorus.

(v) Available potassium.

The methods used are those standardized and tested for a particular area. In addition to the above items, other items, as considered necessary according to the problems, may be analysed.

Based on soil analysis, the soils are classified into categories according to ratings given in Table.

Defining Soil-Test and Fertilizer Recommendations : From the results of analysis of the soil samples, the data contained in the information sheet, and the results of field experiments gathered by the laboratory fertilizer recommendations are made in the proforma.

Rating-chart for the Soil-test Data

Nutrient	*Low*	*Medium*	*High*
Organic carbon (as a measure of available nitrogen)	below 0.5 per cent	0.5-0.75 per cent	Above 0.75 per cent
or			
Available nitrogen (N)	below 280 kg/ha	280-560 kg/ha	About 560 kg/ha
Available phosphorus (P)	below 10 kg/ha	10-25 kg/ha	Above 25 kg/ha
Available potassium (K)	below 110 kg/ha	110-280 kg/ha	Above 280 kg/ha

	pH		
(i) Acidic	pH below 6	(iii) tending to become alkaline	pH 8.6-9
(ii) Normal to saline	pH 6-8.5	(iv) Alkaline	pH above 9

Total soluble salts (conductivity in millimhos/cm)

(i) Below 1—Normal

(ii) 1—2—Critical for germination

(iii) 2-4—Critical for growth of the sensitive crops

(iv) Above 4—Injurious to most crops

Soil-testing Laboratory

(Soil-Test Report and Recommendations)

Sample No.	*Date*
Texture	
pH	Quantity of nutrient to be
E.C. (mmhos/cm 25°C)	applied in kg/ha
Organic carbon (%)	Nitrogen (N)
Available phosphorus	Phosphorus (P_2O_5)
(kg/ha)	
Available potassium (kg/ha)	Potassium (K_2O)
Compost or green manure	Tonnes per ha
Opening of drains	
Flooding	Times
Application of	Kg lime/ha
Application of	Kg gypsum/ha
	In-charge (Soil Chemist)
	of the Laboratory.

If desired, additional information for individual requirements regarding the time and the method of application of fertilizers, precautions to be taken, etc. may also be obtained.

4

Pattern of Cropping

Cropping activities go on all the year-round in India, provided water is available for crops. In northern India, there are two distinct seasons, *kharif* (July to October), and *rabi* (October to March). Crops grown between March and June are known as *zaid.* In some parts of the country, there are no such distinct seasons, but there they have their own classification of seasons. The village revenue officials keep plot-wise record of crops grown in each season. These are annually compiled district-wise, state-wise and on an all-India basis. From these records one could calculate the relative abundance of a crop or a group of crops in a region. These crops are grown sole or mixed (mixed-cropping), or in a definite sequence (rotational cropping).

The land may be occupied by one crop during one season (mono-cropping), or by two crops (double-cropping) which may be grown in a year in sequence. Of late, the trend is even towards more than two crops (multiple-cropping) in a year. These intensive croppings may be done either in sequence or even there may be relay-cropping—one crop undersown in a standing crop. With wide-rowed slow-growing crops, companion crops may be grown. There are various ways of utilizing the land intensively. It is proposed to give a synoptic view of cropping patterns prevalent in the country. Before dealing with the cropping patterns, a brief description of the factors that determine the cropping systems of an individual locality or region are briefly presented here.

In any locality, the prevalent cropping systems are the cumulative results of past and present decisions by individuals, communities or governments and their agencies. These decisions are usually based on experience, tradition, expected profit, personal preferences and resources, social and political pressures and so on. Essentially, they are answers to some of the following questions:

(a) What crops with the present pest-and-disease- control methods are ecologically practicable?

(b) What interactions occur among the ecologically practicable crops, and the chosen crops and must be combined in a special way (rotations) in the farming systems?

(c) Are any of the ecologically feasible crops ruled out by infrastructural factors?

(d) Which of the crops, now remaining on the list, are most profitable (or yield most food in a subsistence agriculture)? In what combinations and at what level of input application would they make the best use of local land, climate and input resources in the short-term and long-term situations bearing in mind the degree of food and income security required by the individual farmer and the community?

(e) What operational factors rule out or amend the size and the method of any of the economically preferable crop combinations thereof?

(f) Finally, are the crop combinations, the farming systems and the input level suggested by this process of the individual farmers compatible with his own skills, enterprise preferences, health, age and capital?

The climatic, edaphic and socio-economic diversity of the Indian crop-production scene is dotted with many cropping patterns. With a geographic area of 328.048 million hectares, stretching between 8°N and 36°N latitude and between 68°E and 98°E longitude, its altitude varying from the mean sea-level to the highest mountain ranges of the world, India presents a range and diversity of climate, flora and fauna, with a few parallels in the world. The country presents the paradox of containing in it the station with the highest mean annual rainfall in the world (Cherrapunji in Assam) and also dry, semi-desert area in Rajasthan.

The variability of rainfall is most important in all states, but especially where rainfall is low. In parts of Rajasthan and the Deccan, the variability is more than 100 per cent of the mean. Years of drought account for only too frequent a history of crop failures, whereas the years of flood also cause very considerable loss of agricultural production. Temperatures also vary greatly, both geographically and seasonally. Northern and central parts of India in January have temperatures comparable with those in Europe in July, though with a greater daily range, but in these places in the pre-monsoon months the maximum temperatures of over 40°C are reached over a large area. Frost may occur in winter even in the plains, as far south as a line drawn through Bihar and Madhya Pradesh and may be heavy in Kashmir and areas north of Punjab.

Socio-economically, the peasantry ranges from the relatively affluent Punjabi farmers who operate with a high-input intensity in agriculture to the subsistent farmers of eastern and central India. They even today, sometimes, practise shifting cultivation. Between these two extremes, various intensities of cultivation are practised. The outstanding fact on the socio-economic front is the smallness of the holdings, the average farm-size in most areas being lower than that is in most tropical countries.

Crop production, therefore, presents such an enormous diversity owing to differences in latitude, altitude and variability of rainfall and edaphic diversity which have been presented in detail in the book. Thus it may not be possible to enumerate and describe here every type of cropping pattern prevalent in the country. Some broad contours of farming, however, emerge. The most important element of farming in India is the production of grains and the dominant food-chain is grain-man. On this basis, the country may be divided broadly into five agricultural regions.

1. The rice region extending from the eastern part to include a very large part of the north-eastern and the south eastern India, with another strip along the western coast.
2. The wheat region, occupying most of the northern, western and central India.
3. The millet-sorghum region, comprising Rajasthan, Madhya Pradesh and the Deccan Plateau in the centre of the Indian Peninsula.
4. The temperate Himalayan region of Kashmir, Himachal Pradesh and Uttar Pradesh and some adjoining areas. Here, potatoes are as important as cereal crops (which are mainly maize and rice), and the tree-fruits form a large part of agricultural production.
5. The plantation-crops region of Assam and the hills of southern India where good-quality tea is produced. There is an important production of high-quality coffee in the hills of the western peninsular India. Rubber is mostly grown in Kerala. Several spices are also grown in Kerala and parts of Karnataka and Tamil Nadu.

> There are some large estates, but most of the growers would come under the category of small holders. Sugarcane, which in many countries is a plantation crop, is almost entirely grown by small holders in India.

There had been a substantial investment in major irrigation works in the colonial days. The post-Independence era saw many multipurpose irrigation works. Lately, interest in the medium and minor irrigation works has increased, especially after the drought of 1966. Thus, at present, an all-India irrigation potential of 38.5 m ha has been created and is expected to increase up to 110 m ha by 2025. Irrigation, especially the minor works, has provided a base for multiple-cropping. The All India Coordinated Crop-Improvement Projects run cooperatively by the Indian Council of Agricultural Research and the agricultural universities have generated short-season, photo-period-insensitive high-yielding varieties of various crops suitable for a high intensity of cropping. The adaptability of these varieties on the farmers' fields has been demonstrated in the National Demonstration Programme spread all over the country. The various developmental and educative programmes, especially the High-Yielding-Varieties Programme, have also resulted in newer cropping patterns involving intensive cropping. The area of rice has increased in Punjab and Haryana. Similarly wheat is now grown in West Bengal and to some extent in the southern states of the country.

All these factors have led to the present cropping patterns, which are getting more and more intensive both in respect of the number of crops grown per year and in respect of the intensity of inputs utilized in the production of these crops.

Kharif Cropping Patterns

As indicated earlier in this chapter, we can hardly describe all the cropping patterns within the framework of this chapter. Therefore only important ones are highlighted. There are many ways in which a cropping pattern can be discussed. A broad picture of the major cropping patterns in India can be presented by taking the major crops into consideration. To begin with, the south-westerly monsoon crops *(kharif)* are considered. They include rice, sorghum *(kharif)*, *bajra*, maize, *ragi*, groundnut and cotton. Among the post-monsoon crops *(rabi)*, wheat, sorghum (*rabi)* and gram can also be considered to be the base crops for describing the cropping patterns.

With such an approach, the crop occupying the highest percentage of the sown area of the region is taken as the base crop and all other possible alternative crops which are sown in the region either as substitutes for the base crop in the same season or as the crops which fit in with the rotation in the subsequent season, are considered in the pattern. Also these crops have been identified as associating themselves with a particular type of agro-climate, and certain other minor crops with similar requirements are grouped under one category. For example, wheat, barley and oats, are taken as one category. Similarly, the minor millets (*Paspalum*, *Setaria* and *Panicum* spp.) are grouped with sorghum or *bajra*. Certain other crops, such as the plantation crops and other industrial crops are discussed separately.

Cropping Patterns in Kharif Season

Among the *kharif* crops, rice, *jowar*, *bajra*, maize, groundnut and cotton are the prominent crops to be considered the base crops for describing the *kharif* cropping patterns.

The rice-based cropping patterns. Rice is grown in the high-rainfall area or in areas where supplemental irrigation is available to ensure good yields. If the crop has to depend solely on rainfall, it requires not less than 30 cm per month of rainfall over the entire growing period. However, only 9 per cent of the area in the country comes under this category, and it lies in the eastern parts. Nearly 45 per cent of the total rice area in India receives 30 cm per month of rainfall during at least two months (July and August) of the south-westerly monsoon and much less during other months. In contrast to these parts, the eastern and southern regions comprising Assam, West Bengal, coastal Orissa, coastal Andhra Pradesh, Karnataka (most part), Tamil Nadu and Kerala receive rainfall of 10 to 20 cm per month in four to eight consecutive months, starting earlier or going over later than the south-westerly monsoon months. With supplemental irrigation, 2 or 3 crops are taken in these areas. However, it has been observed that on an all-India basis, nearly 80 per cent of rice is sown during June-Sept., and the rest during the rest of the season. Area-wise, the mono-season belt accounts for 46.4 per cent (comprising Uttar Pradesh, Bihar and Madhya Pradesh), and the multi-season belt occupies 53.6 per cent of the area (comprising Assam, West Bengal, coastal Orissa, coastal Andhra Pradesh, parts of Tamil Nadu, Karnataka and Kerala).

On an all-India basis, about 30 rice-based cropping "patterns have been identified in different states. In the most humid areas of eastern India comprising Tripura, Manipur and Mizoram, rice is the exclusive crop. In Meghalaya, rice is alternated with cotton, vegetable and fruit crops, whereas in Arunachal Pradesh, where rice is not grown exclusively, the alternative crops being maize, small millets and oilseeds. In parts of Assam, West Bengal, Bihar, Orissa and northern coastal districts of Andhra Pradesh, jute forms an important commercial crop alternative to rice.

In West Bengal, besides rice and jute, pulses and maize are grown on a limited scale. In Bihar, rice is grown over 49 per cent (5.3 m ha) of its cropped area (14.2 per cent of the all-India area), whereas pulses, wheat, jute, maize, sugarcane and oilseeds are the alternative crops. In Uttar Pradesh rice is grown on 19 per cent (4.6 m ha) of its cropped area and represents about 12.4 per cent of the all-India area under this crop. Rice is concentrated in the eastern districts of Uttar Pradesh where the alternative crops are pulses, groundnut, sugarcane, *bajra* and *jowar* in the decreasing order of their importance. Tobacco is grown in some districts.

In Orissa, rice is grown on more than 50 per cent of the area, whereas the alternative crops are: pulses, *ragi,* oilseeds, maize, and small millets. In Madhya Pradesh, rice is grown in the Chhattisgarh area on 4.3 m ha (11.7 per cent of the all-India rice area), but the crop suffers because of inadequate rainfall and irrigation. The important alternative crops of this area are: small millets, pulses and groundnut. Wheat is also grown on a limited scale.

In the southern states, namely Andhra Pradesh, Tamil Nadu and Kerala rice is grown in more than one season and mostly under irrigation or under sufficient rainfall. Together, these three states have over 6.0 m ha, representing over 17 per cent of the all-India area under rice. Important, alternative crops in Andhra Pradesh are: pulses, groundnut, *jowar,* maize, sugarcane and tobacco. In Karnataka, the crops alternative to rice are: *ragi,* plantation crops, *bajra,* cotton, groundnut, *jowar,* and maize. In Kerala, plantation crops and tapioca form the main crops alternative to rice. In Maharashtra, rice is grown mostly in the Konkan area over 1.3 m ha, along with *ragi,* pulses, *rabi jowar,* groundnut, sugarcane and oilseeds. In other states, namely Gujarat, Jammu and Kashmir, Rajasthan and Himachal Pradesh, rice forms a minor crop and is mostly grown with irrigation. However, in Punjab and Haryana and to some extent in western Uttar Pradesh owing to high water-table during the monsoon season, rice has become a major crop on such areas.

The *kharif* cereals other than rice. Maize, *jowar* and *bajra* form the main *kharif* cereals, whereas *ragi* and small millets come next and are grown on a limited area. By and large, maize is a crop grown commonly in high-rainfall areas, or on soils with a better capacity for retaining moisture, but with good drainage. Next comes *jowar* in the medium-rainfall regions whereas *bajra* has been the main crop in areas with low or less dependable rainfall and on light-textured soils. The extent of the area under these crops during the south-westerly monsoon season is maize, 8.6 m ha; *jowar (kharif)*, 11 m ha, and *bajra*, 12.4 m ha. Even though these crops are spread all over the western, northern and southern India, the regions of these cropping patterns are demarcated well to the west of 80° longitude (except that of maize). *Ragi* as a *kharif* cereal (2.4 m ha) is mainly concentrated in Karnataka, Tamil Nadu and Andhra Pradesh which account for more than 60 per cent of the total area under this crop in India. The cropping patterns based on each of these *kharif* cereals are discussed.

The maize-based cropping patterns. The largest area under the *kharif* maize is in Uttar Pradesh (1.4 m ha), followed by Bihar (0.96 m ha), Rajasthan (0.78 m ha), Madhya Pradesh (0.58 m ha) and Punjab (0.52 m ha). In four states, namely Gujarat, Jammu and Kashmir, Himachal Pradesh and Andhra Pradesh, the area under maize ranges from 0.24 to 0.28 m ha in each, whereas other states have much less area under it. Taking the rainfall of the maize-growing areas into consideration, over 72 per cent of the areas receive 20-30 cm per month of rainfall in at least two months or more during the south-westerly monsoon season.

On the all-India basis, about 12 cropping patterns have been identified. They have maize as the base crop. In the maize-growing areas of Uttar Pradesh and Bihar, rice in *kharif* and wheat in *rabi* are the main alternative crops. In some areas, *bajra*, groundnut, sugarcane, *ragi* and pulses are taken as alternative crops. In Rajasthan, maize is grown as an exclusive crop in some areas, whereas at other places, it is replaced by small millets, pulses, groundnut and wheat *(rabi)*. In some areas, *bajra* and rice are also grown as alternative crops. In Madhya Pradesh, mainly the *kharif jowar* is replaced by maize, whereas rice and groundnut are also grown to a limited extent. In Punjab, maize has groundnut, fodder crops, and wheat *(rabi)* as alternative crops. In other states, e.g. Gujarat, rice, groundnut, cotton and wheat form the alternative crops in the maize-growing areas. Rice and wheat are the main alternative crops in the maize-growing areas of Himachal Pradesh, whereas in Andhra Pradesh, rice, *kharif jowar*, and oilseeds are grown in these areas.

The Kharif jowar-based cropping patterns. The area under the *kharif jowar* in India is highest in Maharashtra (2.5 m ha), closely followed by Madhya Pradesh (2.3 m ha), whereas in each of the states of Rajasthan, Andhra Pradesh, Karnataka and Gujarat, the area under this crop is between 1.0 and 1.4 m ha. *Jowar* is mainly grown where rainfall distribution ranges from 10-20 cm per month at least for 3 to 4 months of the south-westerly monsoon or is still more abundant.

On the all-India basis, about 17 major cropping patterns have been identified. In them the base crop is the *kharif jowar*. Most of the alternative crops are also of the type which can be grown under medium rainfall.

In Maharashtra, cotton, pulses, groundnut and small millets are sown as alternative crops. In the adjacent states of Madhya Pradesh, besides the above alternative crops, wheat and fodders are sown. In Rajasthan, wheat, cotton, *bajra* and maize are grown in the *kharif-jowar* tract, whereas in Andhra Pradesh, groundnut, cotton, pulses, and oilseeds form the main alternative crops. Besides cotton and groundnut, *ragi* is sown in the *kharif-jowar* tract of Karnataka, whereas in Gujarat, *bajra*, cotton and groundnut are the major alternative crops.

The bajra-based cropping patterns. *Bajra* is a more drought-resistant crop than several other cereal crops and is generally preferred in low-rainfall areas and on light soils. The area under the *bajra* crop in India is about 12.4 m ha and Rajasthan (4.6 m ha) shares about 2/3 of the total area. Maharashtra,

Gujarat and Uttar Pradesh together have over 4.6 m ha, constituting an additional 1/3 area under *bajra*, in India. Over 66 per cent of this crop is grown in areas receiving 10-20 cm per month of rainfall, extending over 1 to 4 months of the south-westerly monsoon.

On the all-India basis, about 20 major cropping patterns have been identified with *bajra*. However, it may be observed that *jowar* and *bajra* are grown mostly under identical environmental conditions and both have a wide-spectrum adaptability in respect of rainfall, temperature and soil.

Considering the cropping patterns in different states, *bajra* is grown along with pulses, groundnut, oilseeds and *kharif jowar* in Rajasthan. Gujarat has a similar cropping pattern in its *bajra* areas, except that cotton and tobacco are also grown. In Maharashtra, besides having some areas solely under *bajra*, pulses, wheat, *rabi jowar*, groundnut and cotton are substituted for it. In Uttar Pradesh, maize, rice and wheat form the main alternative crops to this crop.

The groundnut-based cropping patterns. Groundnut is sown over an area of about 7.2 m ha, mostly in 5 major groundnut-producing states of Gujarat (24.4 per cent area), Andhra Pradesh (20.2 per cent), Tamil Nadu (13.5 per cent), Maharashtra (12.2 per cent) and Karnataka (12.0 per cent). Five other states, viz. Madhya Pradesh, Uttar Pradesh, Punjab, Rajasthan and Orissa, together have about 17.3 per cent of the total area under this crop. The rainfall in the groundnut area ranges from 20-30 cm per month in one of the monsoon months and much less in other months. In some cases, the rainfall is even less than 10 cm per month during the growth of the crop. The irrigated area under groundnut is very small and that too, in a few states only, viz. Punjab (16.4 per cent), Tamil Nadu (13.3 percent) and Andhra Pradesh (12.5 per cent).

On the all-India level, about 9 cropping patterns have been identified with this crop. In Gujarat, besides the sole crop of groundnut in some areas, *bajra*, is the major alternative crop, whereas the *kharif jowar*, cotton and pulses are also grown in this tract. In Andhra Pradesh and Tamil Nadu, this crop receives irrigation in some areas and rice forms an alternative crop. Under rainfed conditions, *bajra*, *kharif jowar*, small millets, cotton and pulses are grown as alternative crops. In Maharashtra, both the *kharif* and *rabi jowar* and small millets are important alternative crops. In Karnataka also, *jowar* is the major alternative crop, whereas cotton, tobacco, sugarcane and wheat are also grown in this tract.

The cotton-based cropping patterns. Cotton is grown over 7.6 m ha in India. Maharashtra shares 36 per cent (2.8 m ha), followed by Gujarat with 21 per cent (1.6 m ha), Karnataka with 13 per cent (1 m ha) and Madhya Pradesh with 9 per cent (0.6 m ha) of the area. Together, these four states account for about 80 per cent of the area under cotton. Other cotton-growing states with smaller areas are Punjab, with 5 per cent (0.4 m ha), Andhra Pradesh and Tamil Nadu each with 4 per cent (0.31 m ha), Haryana and Rajasthan with 3 per cent each (0.2 m ha each). Most of the cotton areas in the country are under the high to medium rainfall zone.

The cotton grown in Madhya Pradesh, Maharashtra, Karnataka and Andhra Pradesh (4.8 m ha) is rainfed, whereas in Gujarat and Tamil Nadu (1.93 m ha) it receives partial irrigation (16 to 20 per cent of the area). The area under cotton in Punjab, Haryana, Rajasthan and Uttar Pradesh (0.89 m ha.) gets adequate irrigation, ranging from 71 to 97 per cent of the area. These growing conditions, together with the species of cotton grown, determine the duration of the crop which may vary from about 5 to 9 months.

On the all-India basis, about 16 broad cropping patterns have been identified. In Maharashtra, Madhya Pradesh, Andhra Pradesh, and Karnataka, the cropping patterns in the cotton-growing areas are mostly similar owing to identical rainfall.

These patterns include *jowar (kharif* and *rabi),* groundnut and small millets. Pulses and wheat are also grown on a limited area. In some pockets, where irrigation is available, rice and sugarcane are also grown. In Gujarat, rice, tobacco and maize are grown, besides the rainfed crops, e.g. *jowar* and *bajra.*

Pattern of Cropping in Rabi Period

Among the *rabi* crops, wheat, together with barley and oats, *jowar* and gram, are the main base crops among the *rabi* cropping patterns. Generally, wheat and gram are concentrated in the subtropical region in northern India, whereas the *rabi* sorghum is grown mostly in the Deccan. The extent of these areas in different states is as follows:

Crop	*Area*	*Region (percent of the all-India area)*
Wheat	19 m ha	Uttar Pradesh (33), Madhya Pradesh (19), Punjab (13), Rajasthan (8), Bihar (7.5), Haryana (6.5), Maharashtra (5.5)
Gram	8.0 m ha	Uttar Pradesh (25.2), Madhya Pradesh (21.4), Rajasthan (20.7), Haryana (14.2)
Jowar (rabi)	6.5 m ha	Maharashtra (55.0), Karnataka (25.7), Andhra Pradesh (24.3)

The wheat-and-gram-based cropping patterns. These two crops are grown under identical climate and can often be substituted for each other. The core of the wheat region responsible for 70 per cent of the area and 76 per cent of production comprises Punjab, Haryana, Delhi, Uttar Pradesh, and Madhya Pradesh, flanked by Rajasthan and Gujarat in the west region and Bihar and West Bengal in the east region. This area has an extensive irrigation system ranging from 85 per cent area in Punjab to 51 per cent in Bihar. The rainfall during the south-westerly monsoon is also fairly high with over 20 cm to 30 cm of rainfall in at least two out of the four months of the rainy season. However, winter showers are scattered and form loss than 2.5 cm in each month from November to February. On the all-India level, about 19 cropping patterns have been identified with wheat and 7 cropping patterns with gram. In Uttar Pradesh, maize, rice, *jowar* (K), small millets and groundnut form the main crops preceding wheat and gram. Generally, gram is grown on more moisture-retentive soils, but with little irrigation or in areas with less of rainfall. In Madhya Pradesh, wheat is grown with stored moisture, with little irrigation and rainfall during the crop period.

The crop suffers heavily for want of adequate moisture with the resultant low yields (57 per cent of the all-India yield). The *kharif jowar*, groundnut, oilseeds, cotton, small millets and fodders form the alternative crops to wheat and gram. In Punjab, 85 per cent of the wheat area is under irrigation and, therefore, has rice, maize, fodders, *bajra* and cotton as the crops preceding wheat. The area under gram in Punjab is very meagre (4.2 per cent of the all-India gram area). In Rajasthan, the *kharif jowar* fodders and *bajra* precede wheat, whereas gram and other oilseeds form alternative crops in winter. In Bihar, rice, maize and pulses are the main preceding crops, wheat, in the wheat-growing areas, whereas oilseeds and *bajra* are also grown as alternative crops.

In Haryana, wheat and gram are the main alternative crops in winter. Rice, *maize, bajra* and *jowar* form the main preceding crops. In Maharashtra, most of the wheat crop is grown on residual moisture, *bajra* and other small millets or short-duration pulses form the monsoon crop in the wheat areas. Generally, heavy black cotton soils of Maharashtra and the adjacent Madhya Pradesh are left fallow in the *kharif* season for operational difficulties and wheat is grown after the cessation of rains with stored moisture. In Maharashtra, the *rabi jowar* is a crop alternative to wheat.

Rabi jowar-based cropping patterns. On the all-India level, about 13 cropping patterns have been identified with the *rabi jowar*. Maharashtra has the largest number of these cropping patterns, wherein starting with the exclusive *rabi jowar, bajra,* pulses, oilseeds and tobacco are grown as alternative crops. In Karnataka, small millets, groundnut, *bajra,* pulses and oilseeds form alternative crops to the *rabi jowar*. Cotton and tobacco are also grown in some parts of the *rabi-jowar* area of Karnataka. In Andhra Pradesh, short duration pulses, small millets, paddy and oilseeds form the main alternative crops in the *jowar* area.

Process of Crop Planting

Crops under this category include sugarcane, tobacco, potato, jute, tea, coffee, coconut, rubber and other crops, such as spices and condiments. Some of them are seasonal, some annual and some perennial. Generally, the areas occupied by them are very limited as compared with food and other crops. Nevertheless, they are important commercially. Most of them require specific environmental conditions and from the point of view of cropping patterns, they are concentrated in some particular regions. Besides, certain horticultural crops, such as apple, mango and citrus, are important. The extent of the area and the regions in which they are grown are shown below:

Crops	*Area*	*Region (per cent of the all-India area)*
Sugarcane	2.5 m ha	Uttar Pradesh (51), Haryana (6), Bihar (6), Punjab (6), Maharashtra (8), Andhra Pradesh (5), Tamil Nadu (5),
Tobacco	0.427 m ha	Karnataka (3) Andhra Pradesh (48), Gujarat (19.5), Karnataka (8.7), Maharashtra (3.5), Tamil Nadu (3.5),
Potato	0.491 m ha	Uttar Pradesh (33.6), Bihar (20.4), West Bengal (13.3), Assam (5.2), Orissa (4.8)

Contd...

Crops	*Area*	*Region (per cent of the all-India area)*
Jute	0.778 m ha	West Bengal (60), North Eastern Region (18.7), Bihar (17.6), Orissa (6.1), Uttar Pradesh (1.7)
Coconut	1.05 m ha	Kerala (68.3), Karnataka (12.4), Tamil Nadu(9.7) Andhra Pradesh (3.5)
Rubber	0.197 m ha	Kerala (92.8), Tamil Nadu (5.0), Karnataka (1.9)
Cashew	0.264 m ha	Kerala (67.4), Karnataka (12.1), Andhra Pradesh (10.8), Tamil Nadu (9.8), Maharashtra (4.8),
Tea	0.35 m ha	West Bengal, Assam and Tripura (77), Kerala, Tamil Nadu and Karnataka (20),
Coffee	0.138 m ha	Kerala, Tamil Nadu and Karnataka (99)
All fruit-crops	1.8 m ha	Spread all-over India
Onion	0.16 m ha	Maharashtra (18.5), Karnataka (11.7), Andhra Pradesh (12.8), Tamil Nadu (11.2), West Bengal (7.6), Madhya Pradesh (7.2), Orissa (6.8), Punjab (6.2)
Chillies	0.733 m ha	Andhra Pradesh (26.9), Maharashtra (20.4), Karnataka (14.5), Tamil Nadu (10.1), Madhya Pradesh (5.5)
Coriander	0.283 m ha	Andhra Pradesh (36), Rajasthan (23.6), Madhya Pradesh (11.1), Tamil Nadu (10.0)

In several sugarcane-growing areas, mono-cropping is practised, and during the interval between the crops, short-duration seasonal crops are grown. In U.P., Bihar, Punjab and Haryana, wheat and maize are the rotation crops. Rice is also grown in some areas In the southern states, namely Tamil Nadu, Andhra Pradesh and Karnataka, *ragi,* rice and pulses are grown along with sugarcane. In Maharashtra, pulses, *jowar* and cotton are grown.

In the potato-growing region, maize, pulses and wheat are the alternative crops. In the tobacco-growing areas, depending on the season and the type of tobacco, *jowar,* oilseeds and maize are grown in rotation. In the jute growing areas, rice is the usual alternative crop.

In the case of plantation crops, intercropping with pulses and fodder crops is common. Spices and condiments are generally grown on fertile soils. Chillies are rotated with *jowar,* whereas onion, coriander, turmeric and ginger are grown as mixed crops with other seasonal crops.

The Mixed Cropping : Crop mixtures are widely grown, especially during the *kharif* season. Pulses and some oilseeds are grown with maize, *jowar* and *bajra.* Lowland rice is invariably grown unmixed, but in the case of upland rice, several mixtures are prevalent in eastern Uttar Pradesh, with Chhotanagpur Division of Bihar and in the Chhattisgarh Division of Madhya Pradesh. During the *rabi* season, especially in the unirrigated areas of the north, wheat and barley and wheat and gram or wheat + barley + gram are the mixtures of grain crops. *Brassica,* and safflower are grown mixed with gram or even with wheat. Mixed cropping was considered by researchers a primitive practice, but now many researchers regard mixed cropping as the most efficient way of using land. Several new mixtures have recently been suggested. They ensure an efficient utilization of sunshine and land. Breeders are developing plant types in pulses and oilseeds, with good compatibility with row crops.

Prospect of Cropping Patterns: With the increase in population, the irrigated area is increasing and with advances in agricultural science, most of the extensive cropping patterns are giving way to intensive cropping. The development in minor irrigation works has especially provided the farmers with opportunities to crop their land all the year round with high-yielding varieties. This intensive cropping will require an easy and ready availability of balanced fertilizers ana plant protection chemicals and an appropriate price policy for inputs and agricultural produce.

India is a country of small farmers. In the future, the size of the holdings will diminish further. The country has to produce enough for its people without deteriorating the quality of the environment.

This is the challenge of the future for the farmers, agricultural scientists, extension workers and administrators.

5

Health Problems

How to Define : A plant disease is a physiological or structural abnormality that is deleterious to the plant or to any of its parts or products, or reduces their economic value. It expresses itself in a definite series of symptoms.

Nature of Disease : Plant diseases may be temporary or permanent, depending on the speed of recovery. Further, a disease may be localized, involving only a limited part of a plant, or it may be systemic when its infection is generalized and spreads throughout the plant body.

Causes of Disease : Plant diseases may be caused by micro-organisms, including fungi, bacteria, viruses, mycoplasmas, etc. or may be incited by physiological causes including high or low temperatures, lack or excess of soil moisture and aeration, deficiency or excess of plant nutrients, soil acidity or alkalinity, etc. The micro-organisms obtain their food either by breaking down dead plant and animal remains (saprophytes) or by attacking living plants and animals (parasites). In order to obtain nutrients, the parasitic organisms excrete enzymes or toxins and kill the cells of the tissues of the host plant, as a result of which either the whole plant or a part of it is damaged or killed, or considerable disturbance takes place in its normal metabolic processes.

How to Reproduce : Micro-organisms are reproduced and disseminated mainly by spores. Wind, water, seed, diseased plants, cuttings and tubers, animals, men and insects, and soil help to disseminate diseases. Virus diseases are spread as the result of direct contact, through seeds, diseased cuttings and grafts, and more commonly by insect and other vectors. Most of the fungi and bacteria erupt or ooze from the surface after the host tissues have been damaged, whereas the viruses work altogether internally.

Therefore, the control of plant diseases usually lies in preventing infection, as it is very difficult to kill the fungi or bacteria or to inactivate the viruses inside the host plants. However, as the result of recent studies made in the development of systemic fungicides, it appears that in the near future the diseased plants may even be cured. The revolvement of disease-resistant varieties is another important and cheap method of controlling diseases. However, this method of control has its limitations because of the appearance of new races of the pathogens from time to time.

Disease Borne by Seed itself : In the case of seed-borne diseases, the pathogens are carried either on the surface of the seed or within it. When a pathogen is external, it may be destroyed by treating the seed with a chemical, e.g. formalin, TMTD, copper carbonate, Captan, organo-mercurials, e.g. Agrosan and Ceresan. Except Formalin, all chemicals are available in the dust form.

The recommended dust is thoroughly mixed with the seed, preferably in a rotating-drum, so as to uniformly cover the seed surface with a layer of it. While using formalin, dilute it with an equal quantity of water and spray it over a thinly spread layer of grains. The grains are then heaped and kept covered with a tarpaulin overnight. For internally seed-borne diseases, hot-water treatment and solar treatment are very efficacious.

For the treatment of wheat seed with hot water to control loose smut, the seed is soaked in ordinary water for four to five hours and then dipped into hot water at 49° C for one minute and then immersed again into hot water at 54° C for ten minutes. In the case of solar treatment, which is carried out on a bright sunny day during the summer months, the seed is soaked in water for 4 hours from 8 a.m. to 12 noon. It is then taken out and immediately spread in a thin layer on sheets of cloth to dry in the sun. Recently, systemic organic compounds have also been shown to be effective in eradicating both externally and internally seed-borne infections.

Disease Borne by Soil : In the case of soil-borne diseases which attack the seeds or seedlings, the use of soil disinfectants or sterilizers is helpful in destroying the pathogens. The soil treatment involves the use of chemicals formaldehyde, Captan, Thiram, Zineb, organo-mercurials, PCNB, ethylene dibromide, Vapam, etc. They are used to drench the soil, or are broadcast on it, or are applied to it in furrows, or are used to fumigate it.

Disease Born by Air : In the case of diseases caused by air-borne pathogens, the application of chemicals to the foliage is the most effective way of controlling them. Though many fungal and bacterial diseases can be controlled in this way, chemicals have not yet proved very successful in controlling virus diseases. In their case, protective measures have to be aimed at controlling the insect vectors by spraying insecticides on the host plants or by using disease-free material for propagation.

Fungicides in Different Types

Copper, sulphur, organo-mercury and organo-sulphur compounds and antibiotics are some of the important fungicides. Recently, some systemic organic compounds have also been developed. The choice of fungicidal compound depends on the nature of the disease and on the susceptibility of the crop. They are available as dusts or wettable powders and sometimes in liquid form. The method of applying them depends on the availability of water, suitable machinery, the nature of the crop to be treated, the area to be covered and the cost of labour.

Fungicides with Copper : Bordeaux and Burgundy mixtures, and the dispersible forms of cuprous oxide, copper oxychloride, and basic carbonate are among the important copper fungicides. Bordeaux mixture is prepared by mixing solutions of copper sulphate and quicklime in certain proportions. For preparing 5:5:50 Bordeaux mixture, 2 kg of copper sulphate and 2 kg of quicklime are dissolved in 225 litres of water. At first, the stock solutions of the two constituents (1/2 kg in 4 litres of water) are prepared. Copper sulphate is dissolved by placing it in a piece of gunny sacking and then suspending it in water in an earthen or wooden vessel.

Lime is first slaked with a small quantity of water and is made into a thin paste before the remaining water is added to it. The two stock solutions are mixed together just before use by pouring the copper sulphate solution into the lime solution. The final reaction of the mixture should be either neutral or slightly alkaline. The reaction can be tested with litmus paper, or by dipping a bright iron nail into the mixture. The litmus paper must not turn red and the nail must remain bright, i.e. without any copper deposit on it. Bordeaux mixture deteriorates on standing; hence only fresh preparations should be used. However, this deterioration can be prevented appreciably by adding cane-sugar to the mixture at the rate of 60 grammes per 225 litres.

Burgundy mixture is a variation of Bordeaux mixture, in which lime is replaced by sodium carbonate (washing-soda). The method of preparation is just the same as for Bordeaux mixture. The usual strength of the solution is 4—6¼—50. There are a number of proprietary copper fungicides available in the market, e.g. Perenox, Persian, Blitox, Cuprokyt, Cuprosan, Fytolan, which are as efficacious as, and in some cases better than, Bordeaux mixture. The proprietary products also contain dispersing and wetting agents which make them more effective.

Fungicides with Sulphur : The important sulphur fungicides in use are sulphur powder, wettable sulphur, lime-sulphur and organo-sulphur. Sulphur-dust is a very cheap fungicide, but its efficacy depends upon the size of the particles which should pass through a sieve of 250 to 300 meshes per inch. Wettable sulphur is used on crops which are liable to suffer from lime-sulphur scorching. The lime-sulphur solution is prepared as under :

Water (225 litres) is first boiled in a vessel of iron, tin or zinc (but not of copper) and then quicklime (20 kg) and sulphur (55 kg) are added to the boiling water. The mixture is stirred and boiled for one hour. The resultant liquid is strained and its density is determined with a hydrometer ; enough water is added to bring the specific gravity to 1.005 or 1.010 (see table, p. 300) before use. Proprietary products, e.g. Sulfinette and Limsul, are available in powder form. On dilution, they are ready for use as the lime-sulphur spray.

The lime-sulphur spray should not be used on tender foliage, as it is likely to scorch it. A number of organo-sulphur compounds, e.g. thiram and dithiocarbamates, are extremely safe to use on most of the plants with tender and sensitive foliage. Tulisan, Zineb and Ziram are some of the proprietary products, based on organo-sulphur compounds.

Lime-sulphur Dilution Table

Density (Degree Baume)	*Gallons of water to be added to one gallon of concentrated solution to give :* *Sp. gr. 1.005*	*Sp. gr. 1.010*
28	22	46
29	23	48
30	24	50
31	26	52
32	27	54
33	28	57
34	29	59
35	30	61
36	31	63

Fungicides of Other Nature

(a) *The Organo-mercurials* : These have proved to be effective in controlling seed-borne and some soil-borne diseases. They are applied as dusts in extremely small doses to the seed before sowing. They have no harmful effect on germination. Rather, they prevent emergence losses caused by soil-borne pathogens and result in a better stand of the crop. There are a number of organo-mercury preparations, e.g. Agrosan GN, Ceresan, Semesan and Parrugen. They are, however, poisonous to human beings and animals. Hence great care is necessary in handling these fungicides and the seeds treated with them. Some vegetable seeds and other seeds, e.g. those of linseed, are non-tolerant of mercury. In the recent years certain chemicals, such as Captan, Thiram, Busan-72 and Dithane M-45, and some systemic fungicides have been developed. They may be used for seed treatment instead of the mercurials.

(b) *The Systemic Fungicides* : Systemic fungicides can move mainly upwards only within the plant and kill the pathogen away from the point of application. They can be applied as seed or soil treatment, and as sprays. Some of the important systemic fungicides are : oxathin derivatives (Plantvax and Vitavax), Benlate, Demosan and Bavistin.

(c) *The Antibiotics* : Antibiotics are mainly used against bacterial plant pathogens. Some of the promising antibiotics are : streptomycin sulphate, streptocycline, terramycin, agrimycin-100 (a mixture of streptomycin and tetracycline), agrimycin-500 (containing copper), etc.

Steps for Controlling Cultural Barriers : Cultural practices, such as crop rotation, mixed crop ping, adjustment of the date of sowing, depth of sowing and field sanitation through the removal of diseased debris, weeds, alternate and collateral hosts are very effective in controlling plant diseases. A strict enforcement of quarantine, especially in the case of imported plants, helps to prevent the introduction of new diseases into a country.

As virus diseases are difficult to control or cure with chemicals, the use of disease-free propagating material is very important. For this purpose, it is necessary to establish disease-free nurseries at isolated places.

Control measures against some of the important diseases, together with their casual organisms and symptoms, are described in the following pages.

6

Role of Climate

Plant growth and development are primarily governed by the environmental conditions of the soil and climate. The success or failure of farming is intimately related to the prevailing weather conditions. The modification of weather, except on a very limited scale, is yet in the realm of experimentation. It is, nevertheless, possible to optimize farm production by adjusting cropping patterns and agronomic practices to suit the climate of a locality.

Weather assumes significance in nearly every phase of agricultural activity from the preparatory tillage to harvesting and storage. Even after the produce is stored, weather continues to affect the fortunes of the farmers, as the reports of good or bad weather elsewhere tend to upset the price trends.

As weather is the single major limiting factor in crop production successful farming calls for appropriate decisions in the light of weather conditions in the matter of the time of sowing, transplanting, scheduling of irrigation, timing of fertilizer application, using of pesticides, etc.

A sound knowledge of the climatic factors and an understanding of the complex processes of interaction between the climate and the biological processes of the plants are essential to a scientific approach to farming, based on planned cropping patterns and improved land-and-water-management practices.

Weather Conditions

India is a land of many climates and varieties of soils affording scope for much diversity in agriculture. The geographic location and the physical features largely determine the climate of the country.

The lofty Himalayas run along its entire length in the north. To the south of this barrier are the eluvial plains watered by great rivers. In the farther south, there lies the plateau of peninsular India skirted by narrow coastal strips, the Arabian Sea to the west, the Bay of Bengal to the east and the Indian Ocean to the south.

The pattern of rainfall all over India reflects the climatic variations in the different parts of the country. It varies from per-humid in north-east India to arid in Rajasthan. A belt of arid or semi-arid climates extends from the north to the south, dividing the humid climates of the West Coast and the central and eastern parts of the country, where the annual rainfall is generally less than 1,000 mm. This belt is the dry farming tract of India where new crops to augment food production can be introduced.

Precipitation : Rainfall is the most important of the climatic factors. The areas of very heavy rainfall exist on the windward side of the Western Ghats, the Khasi Hills and the Himalayas. These are the source regions for many of the major river systems of the country, particularly the Himalayan ranges. The north-western parts of India are the driest, Rajasthan receiving less than 500 mm of rain annually (Table). A large part of the country receives, on an average, rainfall less than 1,000 mm per annum. Moreover, the average annual rainfall of 1,050 mm is the highest in any part of the world. The irrigation in the semi-arid or arid belt is not sufficient and the water balance is precarious. The agricultural potential of these dry farming tracts can be increased in the near future only by adopting a suitable package of practices aimed at the optimum utilization of available moisture potential through improved soil-and-water management.

South-West Monsoon : With the exception of Jammu and Kashmir in the extreme north and in Tamil Nadu in the south, 80-90 per cent of the rainfall over the country occurs mostly during the south-west monsoon season. The success of agriculture in India, therefore, depends primarily on the timely onset, the proper amount and the distribution of rains in a season. The dates of the onset of the monsoon in different parts of the country and the intensity and the distribution of rain display large variations in time and space and from one year to another.

Normally, The south-west monsoon reaches the Kerala coast by the end of May, advances along the Konkan coast in early June and extends over the entire country by the end of July. The rains continue up to the end of September, when the south-west monsoon recedes. In November and December, the north-east monsoon is the main contributor to the amount of rainfall over the south-eastern portion of the Peninsula.

Depression of Monsoon : The activity of the south-west monsoon is not uniform in time and space during the whole season. The intensity and the distribution of rainfall are controlled by a series of tropical depressions or low-pressure systems which originate near the head of the Bay of Bengal and travel across the country in a west-north-westerly direction. Heavy rainfall occurs mainly to the south of the tracks of these depressions. Three or four depressions form in a month during the monsoon. When in some years they are scarce, the rainfall will be confined to the Western Ghats and the mountain ranges of Assam and the foothills of the Himalayas, with the interior parts of the country not getting their usual share. And there is drought in these areas.

Even when the depressions are of normal frequency and intensity, their tracks determine the distribution of rainfall over northern and central India. When they follow a north-westerly track across the plains of the Ganga, there will be floods in northern India and drought in the Peninsula. The reverse is true when they take a westerly track across central India when the northern states are affected by drought. The conditions in the states, where the tracks of depressions terminate, e.g. in Gujarat and Rajasthan, tend to be erratic. When a depression reaches these states, they get abundant rains; otherwise, they are subject to a prolonged drought.

The monsoon depressions can be said to be the single factor that controls the distribution of rainfall over the whole of India. These, in turn, affect both irrigation and agriculture.

Seasonal Rainfall and Percentage of Annual Rainfall (millimetres)

Meteorological sub-divisions	*Winter (Jan. to Feb.) Rainfall (mm)*	*Percentage of annual*
Bay Islands	81.3	3
Arunachal Pradesh	160.1	4
Assam and Meghalaya	54.0	2
Nagaland, Manipur, Mizoram and Tripura	55.1	2
Sub-Himalayan West Bengal	30.3	1

Contd...

Meteorological sub-divisions	*Winter (Jan. to Feb.) Rainfall (mm)*	*Percentage of annual*
Gangetic West Bengal	38.9	3
Orissa	39.9	3
Bihar Plateau	52.6	4
Bihar Plains	34.0	3
U. P. East	34.7	4
Uttaranchal (Hills)	105.6	6
U.P. West (Plains)	41.4	5
Haryana (including Delhi and Chandigarh)	38.5	7
Punjab	60.4	10
Himachal Pradesh	172.2	11
Jammu and Kashmir	191.7	19
Rajasthan, East	15.0	2
Rajasthan, West	11.3	4
Madhya Pradesh, West	23.2	2
Madhya Pradesh, East	44.4	3
Gujarat	4.4	—
Saurashtra and Kutch	4.1	1
Konkan	3.6	—
Madhya Maharashtra	6.8	1
Marathwada	11.7	2

Contd...

Meteorological sub-divisions	*Winter (Jan. to Feb.) Rainfall (mm)*	*Percentage of annual*
Vidarbha	31.1	3
Coastal Andhra Pradesh	23.7	2
Rayalaseema	15.0	2
Telangana	20.1	2
Tamil Nadu	53.4	5
Coastal Karnataka	4.6	—
Interior Karnataka North	6.0	1
Interior Karnataka South	10.0	1
Kerala	38.4	1
Arabian Sea Islands	44.0	3

Contd.

Summer (March to May)		*Monsoon (June to Sept.)*		*Post-Monsoon (Oct. to Dec.)*		*Annual*
Rainfall (mm)	*Percentage of annual*	*Rainfall (mm)*	*Percentage of annual*	*Rainfall (mm)*	*Percentage of annual*	
447.5	15	1776.5	59	689.2	23	2994.5
958.1	23	2742.8	66	281.1	7	4142.1
673.0	24	1838.0	67	259.2	9	2752.0
549.3	23	1610.1	66	220.1	9	2434.6
482.9	15	2436.4	78	176.6	6	3126.2
177.7	12	1084.5	76	134.2	9	1435.3
127.5	9	1136.9	76	177.9	12	1482.2

Contd...

Summer (March to May) Rainfall (mm)	Percentage of annual	Monsoon (June to Sept.) Rainfall (mm)	Percentage of annual	Post-Monsoon (Oct. to Dec.) Rainfall (mm)	Percentage of annual	Annual
89.7	7	1126.1	82	103.6	7	1372.0
74.3	6	1022.9	85	71.7	6	1202.9
30.4	3	889.1	88	53.5	5	1007.7
102.6	6	1378.8	84	62.4	4	1649.2
30.6	4	722.7	87	36.1	4	830.8
30.6	6	449.5	83	20.6	4	539.2
52.1	9	461.0	76	28.3	5	601.9
195.5	12	1156.5	72	78.5	5	1602.6
237.9	24	472.3	48	92.7	9	994.6
16.4	2	651.6	93	21.1	3	704.1
15.1	5	277.4	89	7.6	2	311.4
21.3	2	947.2	91	53.2	5	1044.9
49.6	4	1228.5	87	79.2	6	1401.7
10.6	1	929.6	96	31.9	3	976.5
10.6	2	449.6	93	18.3	4	482.6
36.1	1	2696.4	94	135.9	5	2872.0
38.4	4	769.6	84	105.9	11	920.7
31.4	4	644.8	83	85.7	11	773.6
34.7	3	956.7	87	77.1	7	1099.6
88.1	9	570.3	57	326.2	32	1008.3
76.9	11	368.8	55	217.1	32	677.8

Contd...

Summer (March to May) Rainfall (mm)	Summer Percentage of annual	Monsoon (June to Sept.) Rainfall (mm)	Monsoon Percentage of annual	Post-Monsoon (Oct. to Dec.) Rainfall (mm)	Post-Monsoon Percentage of annual	Annual
56.9	6	756.5	82	93.0	10	926.5
147.1	15	332.3	33	475.3	47	1008.1
144.5	4	2861.9	88	253.8	8	3264.8
85.5	13	440.6	65	142.9	21	675.0
163.0	13	841.5	67	230.4	19	1244.9
406.1	14	2003.1	67	548.5	18	2996.1
203.2	13	987.2	63	338.0	21	1572.4

Monsoon Breaking : The activity of the monsoon is brought to a halt for a week or two almost all over the country, occasionally during the monsoon. These are called 'breaks' in the monsoon and may occur in any of the monsoon months. The more prolonged breaks are likely during the mid-monsoon month, i.e. August. When a long break occurs during July at the beginning of the cropping season, its effect on the growing crops is quite harmful. The break in the monsoon is generally heralded by the shift of the monsoon trough from the plains of northern India to the hills. Simultaneously, a few low-pressure waves move westwards from the southern part of the Bay of Bengal across the southern part of the Peninsula. Consequently, rains are confined to the hills and the submontane districts of the eastern Himalayas and to the extreme south of the Peninsula. Owing to very heavy rains in the catchments of the Brahmaputra and its tributaries and in the catchments of some of the rivers of northern Bengal and Bihar, floods in these rivers are possible during the 'breaks' season. For the rest of the country, it is a rainless period and may occasionally lead to disastrous consequences owing to crop failures over wide areas.

Storm Due to Thunder and Hail : Generally before and after the monsoon season, thunderstorms with associated showers of rain and hail are the predominant weather phenomena. When sufficient moisture is not present in the air, as over north-west India, only a dust-storm may result. Thunderstorms, with associated squalls, are of short duration, but some are very violent and destructive like the *Kalbaishakis* of Bengal. Dust-storms occur over north-west India. Some violent thunderstorms are accompanied with hail, especially in northern and central India and occasionally in the interior of peninsular India. The frequency of hailstorms is small in winter, but increases generally as the season advances to summer. During the monsoon season, hailstorms are practically absent from the whole country.

The number of days with hail is about 6-7 per year over Himachal Pradesh and its neighbourhood, but it decreases to one in two years over the adjoining plains. Over Bengal, Bihar, Uttar Pradesh and Madhya Pradesh, hailstorms occur about once a year. Hailstorms are comparatively rare over the coastal tracts of the Peninsula.

In central India, especially in February and March, hail-storms occur on one or two occasions in the year, and pose a potential hazard to *rabi* crops. Protection against violent hailstorms is possible by adopting artificial hail-suppression methods. It is difficult to forestall a hailstorm, since its occurrence is sporadic and confined to very limited areas in a thunder-storm which in itself is a highly localized phenomenon.

Storm Due to Cyclone : Cyclones from the Bay of Bengal and the Arabian Sea ravage the Indian coasts during the pre-monsoon period from April to June and during the post-monsoon period from October to December. On an average 2-3 storms may be expected in a year. On the coast, many storms cause widespread damage to the standing crops owing to high winds and the associated tidal waves, which are responsible for about 90 per cent of the deaths in low-lying areas. With the aid of modern tools for detecting and tracking storms, such as the radar and meteorological satellites, the Indian Meteorological Department is now in a position to give advance warnings of impending cyclones with sufficient accuracy. A chain of eight powerful cyclone-warning radars is to be installed in the Fourth Plan, of which two have already been installed at Vishakhapatnam and Madras, and two more are to be installed at Calcutta and Paradeep by the middle of 1973. Weather-satellite-monitoring stations are operating at 5 centres of the IMD, including the three cyclone-warning centres.

Disturbances in Western Part : During November to May, cloudy weather and light rainfall in the plains and snowfall in the hills in the north occur in association with a series of disturbances that enter India through Iran, Afghanistan and Pakistan and move eastwards across northern India and Assam. The average frequency of these western disturbances is two in November and May and four or five from December to April. These disturbances result in light rains which benefit the *rabi* crops. In the winter months, the disturbances are also followed by cold waves over northern and central India. These cold waves are sometimes injurious to vegetable and fruit crops.

Differences in Precipitation

The actual rainfall in any year in a sub-division may differ widely from the normal, as the quantum of precipitation depends upon the nature and the intensity of synoptic weather systems. Areas with low or scanty rainfall are generally the areas of high variability in rainfall. In parts of Saurashtra and Kutch the coefficient of variation, $\left(\text{i.e.} \frac{\text{standard deviation}}{\text{mean}} \times 100\right)$ is 40-50 per cent, whereas in western Rajasthan, it can be as high as 60-80 per cent. The years of deficient rainfall and the years of excessive rainfall occur with no recognizable periodicity. Further, it is not unusual that, in some years, whereas some parts of the country are in the grip of drought, some others are paradoxically ravaged by floods.

Drought. Food production in India being marginal, drought poses many problems. Irrigation facilities available in this country are also limited and, therefore, when droughts occur, they cause partial or complete crop failure. If failures occur in consecutive years, it becomes a national calamity, putting great strain on the economy of the country.

According to the practice followed by the Indian Meteorological Department, drought is taken to have occurred over an area where the annual rainfall is less than 75 per cent of the normal. When the annual rainfall is less than 50 per cent of the normal, it is called severe drought. Drought can occur anywhere in India, depending upon the distribution of rainfall.

Attempts to quantify drought in the form of an index have been made by using various techniques. Pilot studies have been conducted in India, with special reference to Bihar State, in order to arrive at a rational definition. The Palmer drought index, which takes into account rainfall, evapotranspiration and soil moisture is considered a comprehensive approach to this problem. The computation of the Palmer drought index has been carried out for the different sub-divisions of the country. The computations show that, on the average, drought is experienced on 20-25 per cent of the days in each of the months of the *kharif* season over large areas of the country.

Another aspect of great interest in respect of drought is whether there is any periodicity in its occurrence. The power-spectrum analysis of the rainfall series and that of the Palmer drought index series show some relation to the quasi-biennial and the eleven-year sunspot cycles in some areas. The amplitude of the cycle is, however, too small to exert a significant influence.

Connected with the above points is also the problem whether the rainfall series over an area exhibits any trend, cyclic variation or persistence. A number of studies have been carried out on this aspect over different regions of the country. By and large, over the areas studied, it has been found that the rainfall series is found to be of a random nature with no significant trend. Hence it would not be feasible statistically to predict the amounts of rainfall during any particular year from a knowledge of the past rainfall occurrences alone. The probabilities of the rainfall occurrences can, however, be used to estimate the risk of poor rainfall and the attendant drought.

Flooded Area : Many parts of the country, particularly the areas drained by large river systems, suffer from devastating floods. Widespread floods occurred in 1878, 1892, and 1917 in the then many provinces of India, and again in many states of the country in 1954. Rainfall, exceeding 40-50 cm in 24 hours, occurs occasionally along the west coast, in the coastal districts of Tamil Nadu and in southern Andhra Pradesh, in Assam, West Bengal and in the foothills of the Himalayas.

Some of the long-term measures to minimize the damage because of floods consist in the afforestation of the upper catchment areas of the rivers, the construction of river embankments, and the execution of multi-purpose river valley projects. Individual farmers can lessen the damage to their crops owing to heavy rains or floods by making the necessary drainage channels in their fields. Plans are in hand for forecasting floods and for arranging expeditious facilities to take timely action to save the standing crops from devastation by floods. A flood meteorological organization has been drawn up for the country and is under implementation during the current and the subsequent plan periods.

Evapotranspiration and Water Balance : Apart from severe drought leading to crop failures, dry spells of varying lengths occur in the monsoon season. They cause water stress in plants, and the growth and yields are adversely affected, if such stress periods occur during the critical phases of the growth of the crops. The extent to which the water needs of crops are met in a locality depends on the rainfall input, the losses due to runoff and the evapo-transpiration-evaporation from the soil surface and transpiration by the plants. It is important in water-balance studies to know the extent of potential demands on water through evapotranspiration. The evaporation from a free water surface and the potential evapotranspiration depend upon the radiation—incoming as well as outgoing—and the evaporative power of the atmosphere, which again is largely dependent on the saturation deficit and the air movement. In the case of potential evapotranspiration, the extra-roughness of the vegetation cover in relation to water surface is also considered. The areas of high annual potential evapotranspiration are the extreme west of Rajasthan (Jaisalmer) and the extreme south of Tamil Nadu (Tuticorin).

In large parts of the country where rainfall is meagre, a better exploitation of irrigation potential through major and minor projects, the intensive use of groundwater resources and a more efficient conservation and management of water are necessary to stabilize agricultural yields. The growing of drought-resistant varieties in low-rainfall areas, the methods of soil management and cropping that reduce runoff and increase the storage of soil moisture, such as strip-cropping, contour-bunding, mulches and wind breaks, are some of the ameliorative measures that can profitably be adopted.

Air Temperature : The air temperature is a decisive factor in plant growth, especially in the subtropics. The growing season in these latitudes, where frost conditions set a limit, is defined on the basis of air temperature. In India, high temperatures are also of special importance, as they may have damaging effects on crops during the periods of water stress in the arid or semi-arid regions. Each crop is known to have its own optimum, maximum and minimum temperature conditions. Air temperatures bear a close relationship with the incoming solar radiation and the energy available for the growth of plants.

The maximum temperatures in the country are high in March, April and May over northern Deccan and central parts of India . Later in June and July, the highest values shift northwards and north-westwards. In July and August .Tamil Nadu also experiences high maximum temperatures.

Cold Winds and Frosts : The cold season begins over India by early December. During the winter months, under the influence of western disturbances, which are followed by cold dry northernly winds the night temperatures fall rapidly by 7°-8°C below the normal and sometimes more in regions north of the Deccan Plateau and by about 5°C in the peninsular parts of the country. The ground temperatures are usually much lower than air temperatures. The crops may be subject to sub-zero temperatures and damage due to chilling or ground frost is likely to take place. The frost hazard is greatest in northern Punjab, being 10-20 days each in December, January and February. Southwards and eastwards of this area, frost occurrence decreases rapidly.

Cold waves and frost warnings are issued by the regional and state forecasting offices of the Meteorological Department to enable the farmers to adopt timely protective measures. Irrigation or spraying with water helps to mitigate frost damage in the fields. The use of hectors in orchards will help to raise the temperature by as much as 5°C. Valleys in which cold air stagnates are subject to frost to a greater extent, especially in the grape fields of Madhya Pradesh and Maharashtra. While laying out orchards and tea-gardens, such valleys should be avoided. Trees painted as shelter belts or as shade-trees protect the crops during the periods of mild frost. When the standard screen temperature is 5°-7°C. some plant injury is possible even if the stage of ground frost is not reached. Valuable crops can also be protected by covering the ground with suitable material, e.g. paper, plastic or cellophane, to reduce the cooling of the ground through radiation.

Sun Light : Sunshine is not a limiting factor anywhere in India. Even during the monsoon, it is not the lack of sunshine but excessive rain that is responsible for an occasional set-back to plant growth. However, continued cloudiness during flowering is injurious to all plants. It also creates conditions favourable to the multiplication of pests and diseases. High humidity and warm temperatures are conducive to most plant diseases.

Flow of Wind : High winds or squalls are experienced in association with cyclones, depressions and dust-storms and thunderstorms. High winds cause mechanical damage to crops, such as breaking or lodging. Light winds blowing, from colder parts will bring about a drop in temperature and create frosty conditions. Warm and dry winds cause greater losses through evaporation and result in water stress. Windbreaks planted across the prevailing wind directions afford protection to the plants against damage owing to winds. They are recommended as a useful means to reduce the evapo-transpirative losses, frost damage, and soil erosion by wind.

Crop-yield Combination : Weather elements, e.g. precipitation, temperature and sunshine, influence the yield of crops. In India, weather data are available over long periods for a large network of stations. Yield data for different crops for periods ranging from 25-65 years are also available from different states. With the help of these data, formulae have been evolved to forecast the yield of principal food crops (rice and wheat) subdivision-wise by using regression techniques.

The linear correlation between rainfall and other recorded meteorological parameters and the yield of a particular crop over a large number of years has been worked out for overlapping periods of 7 to 90 days on a very fast electronic computer. The significant periods, thus located, have been used for final regression analysis.

These studies reveal that technology comprising improved

methods of farming, the use of hybrid seeds, fertilizers, etc. have en-chanced the yields in almost all the sub-divisions of the country, significantly in recent years. As regards weather, it had been noticed that whereas the distribution of rainfall during the crop-growing season plays a very important role increasing the yield, the *kharif* rice temperature plays only a minor role. In respect of wheat, the preseason precipitation during September-October and the temperatures during the growing season are the main parameters which control the yield. The in-season precipitation is of less importance.

Such forecasts are expected to help the Government to estimate the production of major crops well before the harvest season. Formulae have been developed for a few major crops over a large part of the country and test forecasts are being issued to the Directorate of Economics and Statistics of the Ministry of Agriculture and Irrigation. Pilot studies to develop similar formulae for other major food crops are also in progress.

Long-range forecasts of weather and crop yields. Indian agriculture mainly depends on the monsoon rains. The farmers anxiously await to the onset of the monsoon and a favourable distribution of rainfall during the rainy season. As there is a considerable variation from year to year in the advent of the monsoon over different divisions, the predicting of the dates of its onset and giving advance indications to the probable distribution of rainfall are of great economic significance.

Since 1886, the Indian Meteorological Department has been issuing long-range or seasonal forecasts for the monsoon rainfall during June-September every year. The method consists in making use of certain weather factors over different parts of the world that have significant relationship with the subsequent monsoon rainfall in India. The first forecast is issued in June for the period from June to September, covering the Peninsula, including Maharashtra, Gujarat, Madhya Pradesh, Telengana, interior Mysore (Karnataka north), the coastal Andhra Pradesh and the coastal Mysore (excluding South Kanara) and north-western India comprising Jammu and Kashmir, western Uttar Pradesh, Haryana, Punjab, Himachal Pradesh, Delhi, Chandigarh and Rajasthan. The second forecast is issued for the latter half of the monsoon season (August-September) for each of the Divisions, the Peninsula and north-western India. The third forecast issued in January is for the probable amount of rain and snow during January to March in northern India. These forecasts are, however, of limited use, since they cover large areas, such as northern India and the Peninsula as a whole. Since the monsoon is a major atmospheric circulation over the globe, a dynamic approach by taking into consideration the global factors is necessary. It is also essential to understand the role of the oceanic picking up of moisture and the underlying oceanic currents in order to enable long-term predictions being made with some degree of confidence.

The Division of Agricultural Meteorology, Poona, prepares every year charts, showing the progress of important weather factors, such as rainfall, temperature and humidity. An outlook on crop situation in India, based on these charts, is issued once a month to the Directorate of Economics and Statistics of the Agriculture and Irrigation Ministry for preparing crop-yield statistics for India.

Agro-climate Zone : Water deficit and water surplus are two parameters of great importance for assessing the irrigation requirements and judging the agricultural potential of any place.

Moisture indices have been worked out for about 230 stations in India and a climatic classification of India has been made.

The potential of agricultural productivity is related basically to the periods when water needs of the crops are met to the optimum extent in the various stages of their growth. When the periods and levels of water availability match with the water needs of the crops, the productivity potential can be expected to be high. The length of the period of water availability, therefore, helps to assess the length of the farming season in a rational manner.

The periods of water availability have been worked out for about 220 stations in India, using (i) the monthly average values of rainfall (R), (ii) the monthly average values of potential evapotranspiration (PE), and (iii) the ground storage of water at the beginning and in the end of each month and the maximum possible ground storage, depending upon the nature of the soil.

The duration-dates of the commencement and cessation of the periods were designated as 'humid' ('H'), 'moist' ('M'), 'moderately dry' ('MD') and 'dry' ('D') corresponding to rainfall greater than, or equal to, PE, PE/2, PE/4 and PE/8, respectively.

Pests and diseases and the use of remote sensing through satellites. The damage caused by pests and diseases to crops is well known. The farmer has to contend with the ravaging effects of locust swarms and diseases, e.g. the potato blight and the cereal rusts. Reduction in the losses caused by pests and diseases by timely and effective control measures will considerably add to food production in the country. The incidence of pests and diseases and their intensity are dependent on certain predisposing weather conditions. A beginning has been made to compile information on the meteorological conditions conducive to locust infestation and migration. Weather conditions predisposing the outbreak and spread of pests and diseases affecting crop yields are being studied in collaboration with the concerned organizations in the states and the information on the incidence and intensity of diseases and pests is being collected.

Locust infestations of various types, viz. egg-laying, hopper emergence, swarms and scattered adults have been observed to be mostly confined to Rajasthan, Gujarat, Haryana and Punjab. When large swarms occur, they spread to Karnataka in the south and Assam in the east, across Madhya Pradesh and Uttar

Pradesh. In recent times, locust activity was maximum during 1950-54, and 1960-62.

The locust swarms in India are exotic in origin and the degree of danger due to them depends on the nature, type and magnitude of infestations farther westwards. Once the swarms enter India, their further migration appears to be associated with the type of prevalent weather systems, the orientation of the axis of the monsoon through the low-level wind flow' and the nature and the duration of precipitation.

The meteorological data are being used in some countries for forecasting the outbreaks of pests and diseases. The correlation between the environmental factors and the rate of development of the pests and pathogens forms the basis of such forecasts. In our country also, a pilot project for a few selected areas is under contemplation for the purpose of issuing forecasts regarding the likely time of the incidence of particular pests or diseases and also for the purpose of control operations, such as the spraying and dusting of pesticides.

When plants are infested with pests and diseases, among other things, chlorophyll is destroyed and there is rapid change in the colour of the plants from green to yellow. Infra-red photographs taken from aircraft and earth-resources satellites would furnish information on the state of crops over large areas, enabling the early detection of plant diseases and the adoption of preventive measures.

Crop Planning in Rainfed Areas : For growing crops that are entirely rainfed and are of short duration, it is desirable to choose a short time interval of a week instead of a month for analysing and presenting rainfall data. Information on the quantum of assured water availability over short intervals is essential for developing farming technology, particularly in dryland areas. Over these areas of scanty rainfall, the distribution of rainfall over short periods is not of the normal type and is highly variable from one year to another. It is made up of a large number of small amounts of rainfall and a very small number of higher amounts of rainfall. The variability increases with the shortness of the interval and the average of weekly rainfall becomes less representative and unmeaningful. In the circumstances, the fitting of the series into a theoretical distribution of the gamma type on the lines of work in the USA and Canada appears to be necessary in India also.

The analysis of weekly totals of rainfall over long periods gives information on the lower limits of rainfall, week by week, that are likely to be exceeded for various percentage probability ranges from 90 per cent to 10 per cent. It is expected that these probability figures will be useful in the identification of periods favourable for sowing different crops and for determining the quantum of moisture that would be available during different phases of crop growth. Again, where the amount of assured moisture in a particular probability range is deficient to meet the needs of a crop in a phase, the extent to which this deficiency has to be made up by irrigation can also be determined.

Weather Modification and Crop Production : Delays in the timely onset of rains and breaks in the rainy spells during the cropping season are the two great problems which the Indian farmer has to confront with, especially those in the dry farming tract. During the monsoon season, in some years, over certain regions the rains abruptly cease. Although clouds persist in those areas, their vertical extent is not large enough to precipitate. On such occasions, the clouds can be made to grow and shed their moisture as rain through appropriate seeding.

The water droplet in a cloud grows a million times its original volume, each around a nucleus present in the atmosphere to become a rain drop. In warm clouds (at temperature above freezing), the growth occurs through coalescence of water droplets and in cold clouds (where the temperatures are below freezing), the growth occurs through the formation of ice crystals. The technique of cloud-seeding aims at correcting the deficiencies of nuclei in the cloud. Silver iodide is used for seeding cold clouds whereas sodium chloride (common salt) is generally used for seeding warm clouds. The seeding is effectively done with an aircraft flying inside the cloud.

From seeding experiments undertaken in various countries, it has been established that there is generally an increase in precipitation, but the quantities of precipitation realized through these experiments vary. In India, cloud-seeding experiments have been conducted at Jaipur, Agra and Delhi in the monsoon seasons during periods ranging from 4-9 years, with the aid of ground-based generators, releasing common salt nuclei into the monsoon clouds. A general increase of 10-20 per cent in precipitation has been observed in these experiments. Similar experiments are planned over other parts of the country. A recent technique of dynamic cumulus modification developed in the USA holds out promise for fruitful trials in our country and efforts are being made to organize these trials in the near future.

Rain-making by cloud-seeding can help to tide over a spell of drought significantly, if carried out successfully during critical periods of crop growth.

Meteorological Service to Agriculture : The Indian Meteorological Department is served by a network of observatories all over the country. There are 484 surface observatories, 54 upper-wind observatories, 19 radio sonde radio-wind stations, 317 hydrometeorological observatories, 123 agromet observatories and about 4,000 rain-gauge stations. They collect information on weather parameters that form the basis on which daily weather forecasts and warnings of adverse weather are prepared for each district by the main forecast centres at Chennai, Mumbai, Nagpur, Delhi and Kolkata. Besides the forecasts of general interest, 'Farmers' Weather Bulletins' (two) also indicate periods of the onset of rain, probable rainfall intensity and duration, breaks in rain, spells of heavy rain and other adverse conditions of weather. These bulletins are broadcast in regional languages by the All-India Radio in their evening rural programmes for villagers. The bulletins are framed after consulting the 'Crop-Weather Calendars' prepared by the Division of Agricultural Meteorology, Poona. These calendars contain information on major crops, the significant weather information district wise in relation to crop phases and the likely adverse weather conditions that affect crops. The calendars are drawn up in consultation with the agricultural departments of various states and can also serve as useful guides for use by growers and extension workers.

Special bulletins are also issued to the districts participating in Intensive Agricultural Development Programme. These bulletins, besides daily forecasts, give a weekly summary of the weather. It is thus possible for progressive farmers to plan agricultural operations so as to take advantage of the favourable weather and mitigate the effects of adverse weather.

Weather forecasts containing pieces of operational advice for farmers will be of great use if they take into account the crop situation and needs in different localities. It is proposed to take advantage of the services of agricultural officials who should collect up-to-date information on crops grown district-wise in various states, their stage of growth and the specific operational advice needed by them in relation to the expected weather conditions over the areas in question. Meteorological and agricultural scientists working jointly in the regional and state meteorological centres would be in a more advantageous position, than hitherto to frame operational weather bulletins of specific significance to the farmers in respect of agricultural operations. The institution of such a service will be planned in the near future.

7

Use of Fertilizers

Defining Soil

Buckman and Brady : "A dynamic natural body on the surface of the earth in which plants grow, composed of mineral and organic materials and living forms."

Hilgard : "Soil is the more or less loose and crumbly part of the outer earth crust in which, by means of their roots, plants may or do find foothold and nourishment as well as all other conditions essential to their growth."

Raman : "Soil is the uppermost weathered layer of the soil earth's crust it consists of rocks that have been more or less changed chemically together with the remains of plants and animals that live on it and in it".

Ordinarily the soil is covering on the solid crust of the earth's land mass.

Soil's Basic Concepts

1. Pedology : The considers soil as a natural entity, a biochemically weathered and synthesised product of

nature and even the origin of the soil, its classification and its description, are involved in pedology.

2. Edaphology : The consider soil as a natural habitat for plant, edaphology is the study of the soil from the stand point of higher plants, even it considers the various properties of soils like-plant growth and production.

Categorization of Soil

The soil classification divided into two parts

(A) Old Classification

- Geological approach : According to this approach, soils were divided into two main groups :
  (1) Sedentary
  (2) Transported
- Though classification was further improved and soils were grouped into red soils, black soils, laterite soils, delta soils, desert soils and tarai soils.

(B) Modern Classification System

(1) Physical classification : According to this system, soil is divided into a number of classes like-sand, clay, loam etc.

- This system is good for agricultural purpose but does not requirements of soil science.

(2) Genetical classification : The classification of soil depends on the genesis or organic of the soil and its subsequent development. This system study based upon the soil profile.

- The soils are classified into well-defined categories, the higher categories are order, sub-order, great soil group and sub group.
- The lower categories are families, series, types and phases.

Marbut's Soil Classification

Soil Order		Sub-order	Soil Group
I. Zonal Soil	(a) Pedalfers	(i) Aretic pedalfers	(i) Tundra soil
		(ii) Temperate pedalfers	(i) Podsol soils
			(ii) Brown and grey podsolic soils
		(iii) Tropical pedalfers	(i) Yellow and red podsolic soils
			(ii) Laterite soils
		(iv) Transition pedalfers	(i) Degraded chernozem soils
	(b) Pedocals	(i) Temperate pedocals	(i) Chernozem soils
			(ii) Chestnut soil
		(ii) Tropical pedocals	(i) Brown soils
			(ii) Desert soils
II. Intrazonal Soil		(i) Halomorphic	(i) Saline and alkaline soils
		(ii) Hydromorphic	(i) Bog soil
			(ii) Meadow soils
		(iii) Calomorphic	(i) Rendizine soils
			(ii) Brown forest soils
III. Azonal Soil		(i) Lithosols	
		(ii) Regosols	
		(iii) Alluvial soils	

Revised Soil Orders

Order		*Soil group*
1. Entisols	—	Azonal soils, alluvial soils, regosols.
2. Vertisols	—	Regurs or black cotton soil.
3. Inceptisols	—	Andosols, brown forest.
4. Aridisols	—	Desert soils, reddish desert, some brown soils.
5. Mollisols	—	Chestnut, chernozem, rendizine, some brown soils.
6. Spodosols	—	Podsols, brown podsolic soils, ground water podsols.
7. Alfisols	—	Grey brown podsolic soils, non-calcic soils, degraded chernozem.
8. Ultisols	—	Red yellow podsolic soils, reddish, brown laterite soils.
9. Oxisols	—	Latosols and laterite soils.
10. Histosols	—	Organic soils, bog soils.

Soils in India

The soils of India have been divided into the following 11 major groups

Red Soils : Red colour in red soils is due to the presence of various oxides of iron. Their main features are a light texture, porous structure, absence of lime and low soluble salts. They are slightly acidic to neutral in reaction, the pH ranging from 6.0 to 7.5.

- Red soils are mostly found in the areas of Madras and Mysore, part of A.P, M.P, Orissa, Bihar, Santhal Pargana (Jharkhand), Mirzapur, Jhansi district of U.P.

Laterite and Lateritic Soils : It is found mostly in the areas of high rainfall. They are light in texture. It is deficient in lime and are slightly to moderately acid in reaction.

- The pH varies from 5.0 to 6.0. They are low in base exchange capacity.
- These soils are found all along the west coast of Maharashtra, Mysore and Kerala, on the tops of hills in the Deccan, M.P.

Black Soils : The soils are dark or dark brown in colour. It is developed from Basaltic rock under semi-arid condition.

- The soils locally known as regur or black cotton soil, deep black soil, medium black soil. The texture ranges from sandy loam to heavy clay.
- One of the characteristic features of black soil is that it swells on wetting during the rainy season and shrinks and cracks in summer season.
- The base exchange capacity of deep black soil is quite high.
- The pH varies from 7.5 to 8.5.
- The soils are deficient in nitrogen and phosphorus and they are rich in potash and lime.

- The black soils are found in parts of Maharashtra, Gujarat, M.P., Rajasthan, U.P., A.P. and Madras.

Alluvial Soils : It is grey or greyish-brown in colour and texture of alluvial soils is sandy loam to clay loam and the structure is also variable, loose and freedraining in the case of sandy soils and compact in the clayey soils.

- These type soils are most fertile.
- Their base exchange capacity is comparatively low and pH varies from 7.0 to 8.0.
- The alluvial soils are found in the areas of Rajasthan, Punjab, U.P, Bihar and W. Bengal.

Desert Soils : It is developed in arid regions. It is mostly sandy.

- Desert soils contain large amounts of soluble salts and varying proportions of lime.
- They have a high pH and are very poor in fertility constituents.
- Desert soils are found in the large parts of Rajasthan, South Punjab and in the range of Kutch.

Saline and Alkaline Soils : Such type of soils are developed in the arid and semi-arid regions. Poor drainage is also responsible for their development.

- These are mainly found in the black soil region in the south and west, in the Indo-Gangtic alluvium in the north and in the deltic regions.

Peaty and Marshy Soils : They are formed in depression under submerged conditions and have a blue colour due to presence of ferrous iron.

- It is found scattered in Kerala, north Bihar and north U.P.

Tarai Soils : Tarai soils have a wet regime. It is foot hill soils and extend in strips of varying widths at the foot of the Himalayas in Jummu and Kashmir, U.P, Bihar and W. Bengal.

Soils under the natural condition are thickly vegetated and swampy.

Brown Hill Soils : The soils are mainly found in the Himalayas on sandstones and shales.

- The surface soils are dark brown loam to silty clay in texture, acidic to neutral in reaction.
- The soil may be classified in order Alfisols.

Sub-montane Soils : These soils are formed in high rainfall regions of sub-Himalayan under coniferous forest. The soils are acidic in reaction.

Mountain Meadow Soils : These soils occur at high elevation in the Himalayas. The soils are shallow with grass vegetation.

Soil Formation

- Soil is formed from weathering of rocks and minerals. Soil formation is slow process.
- The formation of 1 inch soil needs approximately 800-1000 years.

Factors of Soil Formation

The factors of soil formation may be kept under three groups :

(a) Passive factors :

1. Parent material
2. Relief or topography

(b) Active factors :

3. Climate
4. Organisms

(c) Natural factors

5. Time or age of land

(Haryana Public Service Commission (Pre) Exam. 2003)

Soil Profile

The vertical section of the soil showing the various layers from the surface to the unaffected parent material is known as a soil profile.

- The various layers are known as horizons.
- A soil-profile contains three main horizons. They are named as horizon-A, horizon-B and horizon-C.
- The surface soil or that layer of soil at the top is known as horizon-A or the horizon of eluviation.
- The intermediate layer in which the materials leached from horizon-A have been redeposited is known as horizon-B or horizon of illuviation.
- The parent material from which the soil is formed is known as horizon-C.
- Each layer or horizon develops specific morphological features such as the size and shape of particles.

Organic Horizons :

A_0 : Undecomposed organic matter

A_{00} : Partially decomposed organic matter

Alluvial Horizons or A-horizons :

A_1 : Top most mineral horizons, dark coloured horizons, high in organic matter.

A_2 : Light coloured horizons, heavy leaching of clay

A_3 : Transition layer between A and B some times A_3 is absent.

Illuvial Horizons or B-horizons :

B_1 : A transition layer between A and B with properties more nearly like B than A.

B_2 : Organic matter content in higher than that of A_2. Zone of maximum accumulation of clays and hydrous oxide.

B_3 : Transition between B and C with properties more like those of B than those of C below.

C-horizons

C_1 : Partially weathered parent material contains accumulation of lime and gypsum in some soils.

C_2 : Unweathered parent material.

Soil Formation in the Process

Basic Process

1. Addition of water, organic matter.
2. Loses of water, O.M., minerals from soil.
3. Translocation in solution and movement in suspension of clay, O.M.
4. Transformation of minerals and organic matter.

Fundamental Process

1. Humification : Process of decomposition of O.M. leading to the formation of humus.
2. Eluviation : Its means washing out. It refers to movement and removal of material in solution from entire solum.
3. Illuviation : Means washing in it is the process of deposition of soil materials in the lower layers.
4. Horizonation : The development of all the horizons in soil is horizonation.

Pedogenic Process

1. Calcification : It means the deposition of $CaCO_3$ in soil profile.
2. Decalcification : The removal of $CaC0_3$ from soil by leaching.
3. Podzolization : Podzolization is a type of eluviation in which humus and sesquioxides become mobile, leach

out from the upper horizones and are deposited in the lower horizons.

4. Laterization : Laterization is the process that removes silica, instead of sesquioxides from the upper surface or layers and there by leaving sesquioxides to concentrate in the solum.

   - Calcination, podzolization and laterization are zonal soil forming processes where soils produced are under normal prevailing conditions of climate and vegetation.

Components of Soil

The soil consists of 4 major components i.e.,

1. Mineral matter	—	45%
2. Organic matter	—	5%
3. Soil air	—	25%
4. Soil water	—	25%

Mineral Matter : The minerals are variable in size. Some are as large as the smaller rock fragments; others such as colloidal clay particles are so small that they can not be seen without the aid of an electron microscope. It is much effect on the properties of soil.

Organic Matter : Soil organic matter represents partially decayed and partially synthesised plant and animal residues.

- Organic matter improves the physical condition of soils. It also increases water holding capacity.

Soil Air : The soil air contains a number of gases viz., nitrogen, oxygen, carbondioxides and water vapour.

- Pore space between the soil particles is filled partly with soil air and partly with soil water.
- Soil air has effect on the plant and their root growth, even effect on soil micro-organism.

Soil Water : Soil water is the major component of the soil in relation to the plant growth. The water is held within the soil pores.

- Soil water dissolves salts and makes up the soil solution, which is important as a medium for supplying nutrients to growing plants.

Physical Properties of Soil

The physical properties of a soil depends on the amount, size, shape, arrangement and mineral composition of its particles. The physical properties also depend on organic matter content and pore spaces.

- Following are the important physical properties of soil:

1. Soil texture
2. Soil structure
3. Soil density
4. Soil porosity
5. Soil consistence
6. Soil colour
7. Soil temperature

Soil Texture

"Soil texture refers to the relative proportion of particles of various size in a given soil."

Classification of Soil Separates

Soil		*Size in diameter*
1. Coarse sand	—	2.00-0.20 mm
2. Fine sand	—	0.20-0.02 mm
3. Silt	—	0.02-0.002 mm
4. Clay	—	below 0.002 mm

(IAS 2002; U.P.PCS 2003; Himachal PCS 2003)

- Generally soil texture influences soil physical and chemical properties like-water holding capacity, nutrient retention, nutrient fixation, drainage, strength, compressibility, etc.
- Clay, calay loam, silty clay, silty clay loam, clay loam soils are heavy textured soils, in which various crops can be grown these crops are : rice, cotton, sorghum, coriands.
- Loams, silty loams, silt and sandy loams are medium textured soils, in which various important crops grown.
- Sandy, loamy sand, sandy loam, and sandy clay are light texture soils, and these are suitable for potato, tobacco, groundnut and leguminous fodder crops.

Soil Structure

"The arrangement of soil particles and their aggregate into certain defined patterns is called structure". (HAS (Pre) 2003)

- The aggregate formation is influenced by the amount and nature of colloidal clay. The nature of absorbed cations also influenced it. Ca^{+2} and H^{+} bring about better aggregation than Mg^{+2} and K^{+}.
- Sesquioxides i.e., iron and aluminium oxides act as cementing agents for binding sand and silt particles to form aggregates.

Type of Soil Structure : There are four main forms of soil structure

1. Plate-like structure : The horizontal dimensions are much more developed than the vertical When the units are thick, they are called platy, and when thin, laminar.
2. Prism like structure : The vertical axis is more developed than horizontal, giving a pillar-like shape.

- When the top of such a ped is rounded the structure is termed as columnar and when flat prismatic.

3. Block-like structure : All these dimensions are about the same size and the peds are cube-like with flat or rounded faces.
    - When the faces are flat and the edges sharp angular, the structure is called as angular blocky.
    - When the faces and edges are mainly rounded it is called sub angular blocky.
4. Spheroidal structure : All rounded aggregates may be placed in this category, although the term more properly refers to those not over 2 inch in diameter.
    - The aggregates of this group, which are relatively less porous is called as granular.
    - The aggregates of this group which granules are very porous, termed as crumby.

Factors Affecting of Soil Structure : The development of structure of soil is depends on the following factors:

1. Climate	2. Organic matter
3. Tillage	4. Plant roots
5. Soil organism	6. Fertilizers
7. Wetting and drying	8. Temperature
9. Porosity	10. Density
11. Consistence	12. Colour

- Large masses of soil aggregates are called "clods".

Density of Soil

Density represents weight (mass) per unit volume of a substance.

$$\text{Density} = \frac{\text{Mass}}{\text{Volume}}$$

Dividing Soil Density

(i) Particle density: The weight per unit volume of the solid portion of soil is called particle density.

- Generally, the particle density of normal soils is 2.65 gm.
- The particle density is also known as true density.
- Particle density depends on the chemical composition and crystal structure of the mineral particle.
- Particle density is denoted by Dp.
- The Dp of O.M. is 1.1 to 1.4 g/cm^3.

(ii) Bulk density: It is the mass of a unit volume of dry soil including both solids and spaces. The bulk density of general soil is 1.33 g/cm^3.

$$\text{B.D.} = \frac{\text{weight of soil}}{\text{volume of solids and pores}}$$

- The bulk density of a soil is always smaller than its particle density.
- The bulk density of sandy loam is 1.60 g/cm^3.
- The bulk density of caly is 1.10 g/cm^3.
- The bulk density of loam is 1.40 g/cm^3.

Porosity of Soil

Pore Space

- The volume of soil mass that is not occupied by soil particles is known as pore space.
- The pore space is usually occupied by air and water, which is helpful for growth and survival of plants.
- Generally two sizes of pore are found

  (i) Macro pores : These are large pores. Sandy and sandy soils have a large number of macro pores. It is known as non-capillary pores.

(ii) Micro pores : It is known as capillary pores. Clay and clay soils have a large number of micro pores.

Soil Porosity

- Soil porosity is the percentage pore space. Porosity refers to that percentage of soil volume which is occupied by pore space. It can be calculated by the formula :

$$\text{Porosity} = 100 - \left(\frac{\text{Bulk density}}{\text{Particle density}}\right) \times 100$$

Factors Affecting Soil Porosity

1. Soil texture
2. Soil structure
3. Arrangement of soil particle
4. Organic matter
5. Macro-organism
6. Depth of soil
7. Cropping

Soil Consistence

- Soil consistence represents of soil materials to deformation. It depends upon the degree and kind of forces which attract one molecule to another.
- Consistence of soil depends on the texture, colloids, structure and especially the moisture content of soil.
- The consistence of soil is generally described at three soil moisture level.
  (i) Consistence when soil is wet
  (ii) Consistence when soil is moist
  (iii) Consistence when soil is dry

Soil's Colour

- The colour of soil varies widely. It is due to either mineral matter or O.M. and mostly to both for example- red colour is related to unhydrated ferric oxide although maganese dioxide and partially hydrated iron oxide may also contributed to it.

- The yellow colour is due to oxides of iron.
- The soil colour directly influenced soil temperature.
- Brown colour is due to a mixture of the organic matter and iron oxides.
- Black and dark grey colour of soil is due to mainly organic matter.
- White colour of soil is due to silica and lime.
- Some of the bluish and greenish colours are due to the presence of ferrous compound.

Soil Heat : Various Sources

Soil's Temperature

(i) Solar radiation : The main source of soil heat is sun rays, which reach the earth after pass through atmosphere.

(ii) Conduction : Interior of the earth's is very hot, the conduction of this heat to the soil is very slow.

(iii) Some amount of heat is liberated in the chemical and biological process.

- The temperature requirement for optimum crop root development is 15-45°C.
- Black cotton soils absorb about 86% solar radiation.
- Alluvial soils 40% and grass covered soils about 60% of solar radiation.
- The specific heat of soils ranges from 0.20 to 0.23

Impact on Soil Temperature : Elements Responsible

1. Soil texture
2. Soil structure
3. Soil composition
4. Soil colour
5. Soil moisture
6. Slop of the land

7. Vegetative cover
8. Climate
9. Season : During Dec., Jan. month soil temperature are low and temperature is high during summer season.

Soil Water Concept

- Soil water is the major component of the soil in relation to the plant growth.
- Much of the water remains in the soil as thin film. Soil water dissolves mineral salts and make up the soil solution, which is important as medium for supplying nutrients to growing plants.
- The soils hold water due to colloidal properties and aggregating qualities.
- The water is retained in the soil by the various ways like-surface tension, cohesion and adhesion forces, polarity.
- The soil water physically classified into 3 categories

1. *Gravitational Water*

- Gravitational water occupies the larger soil pores and moves down readily under the force of gravity.
- Such water held by a negative tension of 0.3 atm.
- Gravitational water,is not useful for plants.
- It reduces aeration in the soil.
- Undesirable and nutrient leaching take place.

2. *Capillary Water*

- Capillary water is held in the capillary pores.
- Capillary water is retained on the soil particles by surface forces.
- Capillary water held between field capacity and hygroscopic coefficient in microspores.

- Soil water tension varies $\frac{1}{2}$ atm (0.1 atm) to 31 atm.
- It is related to organic matter content and texture of soil.
- The molecules of capillary water are free and mobile and are present in a liquid state.
- It is available water.

3. ***Hygroscopic Water***

- The water held strongly on the surface of soil colloidal particles is known as hygroscopic water.
- Hygroscopic water tension varies from 31 to 10000 atmosphere.
- Hygroscopic water held mostly by soil colloids.
- Hygroscopic water held so tenaciously by soil particles that plants cannot absorb it.
- Unlike capillary water which evaporates easily at atmospheric temperature, hygroscopic water cannot be separated from the soil unless it is heated.

Classifying Soil Water in Biological Terms

1. Available water : The available water is that water which lies between wilting coefficient and field capacity.
   - Even capillary water is known as available water.
   - The water lying between field capacity ($\frac{1}{3}$ atm.) and wilting coefficient (15 atm.).
2. Unavailable water : The unavailable water includes the whole of the hygroscopic water plus a part of the capillary water below the wilting point.
   - Such water held in soil at the permanent wilting point.

3. Super fluous water : The water beyond the field capacity stage is said to be super available water or super fluous water.
    - This water is unavailable for the use of plants.
    - It is also called gravitational water.
    - The moisture tension f ≥ $\frac{1}{3}$ atm.)

Field Capacity

- Sometimes after the application of water in the soil all the gravitational water has drained away, then the wet soil is almost uniformly moist the amount of water held by the soil at this stage is known as field capacity or normal moisture capacity.
- The metric tension is around 0.1 to 0.3 atm.
- At the field capacity water is held with a force of $\frac{1}{3}$ atm.
- Water at field capacity is readily available to plants and micro-organism.

Wilting Coefficient

- As the moisture contents falls, a point is reached when the water is so firmly held by the soil particles that plant roots are unable to draw water. The plant begins to wilt. This stage is termed as wilting point and the percentage amount of water held by the soil at this stage is known as the wilting coefficient.
- Water at wilting coefficient is held with a force of 15 atm.
- The pF at 15 atm is about 4.2 while the pF range of available moisture is 2.54-4.2.
- Moisture with about pF 2.54 is very favourable for plant growth.

Hygroscopic Coefficient :

- The hygroscopic coefficient is the maximum amount of hygroscopic water absorbed by 100 gm of dry soil under standard conditions of humidity and temperature.
- The tension is equal to a force of 31 atm.
- The pF is about 6.0.

Cation Exchange Capacity :

- Cation exchange reaction are very important chemical reactions for the availability of plant nutrients in the soil.
- C.E.C. represents the total cation absorbing capacity of soil.

  Cation exchange in most soils increases with pH. At a very low pH value, C.E.C. is higher and at high pH C.E.C. is lower.
- The cation exchange capacity of humus is 100-300 me/100gm of soil.
- The C.E.C. of montmorilonite is 60-100 me/100gm.
- The C.E.C. of vermiculite is 100-200 me/100gm.
- The C.E.C. of illite is 20-40 me/l00gm.
- The C.E.C. in chlorite is 20-40 me/l00gm.
- The C.E.C. in ascending order
- Kaolinite → illite → montmorilonite → vermiculite → humus

Clay Mineral's Concept

- A mineral is a naturally occurring inorganic surface with a definite chemical composition and distinct physical characters.
- Clay is not amorphous but crystalline and lathe shaped of made up or irregular hexagonal plates.

- Humus is amorpous.
- Commonly clays are mainly made up of silicates of various types and are called silicate clays.
- The other type is formed of aluminum and iron hydroxides, it is called sequioside clay.

Major Silicate Clay Minerals

- The clay particles are composed of individual layers. The most important silicate clay is known as phyllosilicate.
- The silicate clay unit consists of alternate sheets. One type of sheet is dominated by silicon and other by Al and Mg.
- Silica-dominated sheet is called tetrahedral because of its four sided configuration and Al and /or Mg sheet is called octahedral because of eight sided building block.

Categorization of Silicate Clays

- On the basis of number and arrangement of tetrahedral and octahedral sheets, silicate clays are classified into 3 different groups.
  (a) 1:1 type minerals : It means one tetrahedral to one octahedral sheet. Kaolinite is the prominent member of 1 : 1 type group and every hallosite, nacrite and dickite include in this group.

  (Himachal Pradesh PCS (Pre) 2003)

  (b) 2 : 1 type minerals : Octahedral sheet (alumina) is sandwiched between two tetrahedral sheet.
- 2 : 1 type mineral also divided into two parts
  (i) 2: 1 type (swelling type minerals) : Montmorilonite and vermiculite are includes in this group but montmorilonite is prominent in this group.

(ii) 2 : 1 type non-swelling type minerals: In soils, illite is the prominent in this group.

(Himachal Pradesh PCS (Pre) 2004)

- About 20% Si^{4+} of silica sheet is replaced by Al^{3+} which results in high net negative charge in tetrahedral sheet. To satisfy this charge K^+ are strongly attracted in the inter-layer space. Thus K^+ acts as binding agents preventing expansion.

(c) 2 : 2 type minerals : The best example of this group is chlorite. This is basically ferro-magnesium silicates with some aluminium present.

- Chlorite have an extra layer of Mg dominated alumina sheet say brucite [$Mg(OH)_2$].

Acidity of Soil

- Soil acidity is caused by ionisable hydrogen ions, which having less than pH 7.0.
- Soil acidity is common in all regions where precipitation is high enough to leach appreciable quantities of exchangeable base forming cations (calcium and Mg etc.) from the soil layers of the soils.

How it is being Formed ?

- Acid soils are formed under natural conditions as well as artificial by the continuous use of acid-forming fertilizers. Following reasons are responsible for acid soil forming
  1. Minerals (Ca, Mg, etc.) leaching due to heavy rainfall.
  2. Microbial action.
  3. Origin of acid from acid parent rock material.
  4. Removal of bases by crops.
  5. Continuous application of acid-forming fertilizers.
- Low acidity occurs at 6.1-6.5 pH.

- Medium acidity occurs at 5.6-6.0 pH.
- Strong acidity occurs at 5.1 to 5.5 pH.

Negative Impact of Acid Soil on Plant Growth

1. Toxic effects of H^+ ions on root tissues.
2. Disturbance in the balance between basic and acidic constituents through the roots.
3. Adverse effects on the availability of various nutrients i.e., copper, zinc, P_2O_5 etc.
4. Prevalence of plant disease.
5. Soil micro-organism also influenced by acid soil.
6. Due to soil acidity, nutrients such as Ca and K may be deficient.

Reclamation of Acidic Soils

1. Addition of liming material : Soil acidity is the result of the accumulation of a predominance of H ions over OH ions. When lime is added to moist soil, the soil solution becomes charged with calcium ions. These active Ca ions exchange take place with hydrogen ions in the exchange complex.
2. The use of basic fertilizers like sodium nitrate, basic slag etc.
3. Proper soil and water management cheeks leaching of bases.
4. The growing of acid tolerant crops like- sweet potato, oat, castor, rice, barley, maize, turnip etc.

The availability of nutrients at low pH: Al, Fe, Mn, Zn, Cu, Co (all micro elements except B and Mo).

- The availability of nutrients at high pH is Mo, N, K, Ca, Mg, S (all major nutrients except P)
- At the neutral pH -P and B
- Boron available at pH 5.0 -7.0.

Saline and Alkaline Nature of Soil

- Saline and alkaline soils are those soils that have an alkaline reaction or pH is greater than 7.0. This is due to the presence of an excess of sodium salts.
- On the basis of presence of sodium salts the soil divided into two parts :
  1. Saline soils : Saline soils contain an excess of sodium salts.
  2. Alkali soils : In the alkali soils, the exchange complex contains appreciable quantum of exchangeable sodium.
- Alkali soils maybe divided into following groups:
  (i) Saline alkali soils
  (ii) Non-saline alkali soils
  (iii) Degraded alkali soils

Important Points of Saline Soil

1. Saline soils contain usually chloride, sulphate bicarbonates etc.
2. Exchangeable sodium percentage is very low (less than 15%).
3. pH varies between 7.5 to 8.5.
4. EC of soil solution saturated exact is more than 4.0 m mhos/cm.
   - The process of accumulation of salts leading to the formation of soils is known as salinization.

Reclamation of Saline Soil

Mechanical methods

1. Leaching down of the soluble salts.
2. Scraping of the surface soil.

Cultural methods

1. Providing the proper drainage.
2. Proper use of the irrigation water.
3. The use of acidic fertilizers.
4. Use of organic matters.
5. Use of salt free irrigation water.
6. Growing of salt tolerant crops-like barley, cotton; sugarbeet, (high salt tolerant crops), wheat, rice, sorghum, flax etc. (moderate slat tolerant crops).
7. Sowing of seeds in the furrow.

Major Elements of Alkaline Soil

Saline alkali soils

1. Exchangeable sodium percentage is more than **15.**
2. pH varies usually above 8.5.
3. Electrical conductivity is more than 4 m mhos/cm.
4. Generally soluble salts content is more than 0.1%.

Non-saline alkali soils

1. These soils contain Na_2CO_3 (sodium carbonates) in abundance.
2. ESP is more than 15%.
3. pH range varies from 8.5-10.
4. EC is usually less than 4 m mhos/cm.
5. Total salt soluble content is less than 0.1%.
6. These soils are almost barren (usar).

Degraded alkali soils

1. The soil does not contain free $CaCO_3$.
2. pH varies more than 8.5.
3. ESP is more than 15%.

4. EC less than 4 mmhos/cm.
5. It develops columnar structure.

Chemical Methods : Reclamation

(i) Application of gypsum ($CaSO_4 2H_2O$) : By cationic exchange, calcium is often used to replace sodium in alkali soil. If the soil has no reserve of calcium carbonate, the addition of gypsum is essential. When gypsum is used as a reclaiming agent, calcium replaces the exchangeable sodium and converts the clay back into calcium clay.

(ii) Even sulphur used for reclaiming of alkali soil.

(iii) The addition of organic matter increases acidity, which is helpful for lowering pH.

(iv) The use of sulfuric acid.

(v) The use of iron pyrite (FeS_2).

Idea of Manures and Fertilizers

The "manures" word used originally for denoting materials like-cattle manure and other bulky natural manure.

- But later two type manures describe
    1. Organic manure : The organic manures are made up of dead plant and animal remains. The organic manures are farm yard manure, compost and green manure.
    2. Inorganic manures were called "**fertilizers**".

Organic Manures

(a) F.Y.M. : Decomposed mixture of dung and urine of farm animals along with litter and left material from roughages or fodder fed to the cattle.

- F.Y.M. contains N-0.5%, P_2O_5—0.2%, and K_2O—0.5%.

- F.Y.M. is the most commonly used organic manure in India.

Compost

- The process of decomposing organic wastes is called "composting" and the decomposed material is called "compost".
- Ordinary compost can be enriched with N and P though *Azotobacter* and super phosphate respectively.
- When super phosphate is used during compost making it is called super compost.
- When compost prepared by using N-fixing bacteria is called azocompost.
- The compost made from farm waste like sugarcane trash, paddy straw, weeds and other plants and other waste is called farm-compost (0.5% N, 0.15% P_20_5, 0.5% K_2O).

When the compost made from town refuses like night soil, street sweeping and dustbin refuse is called town compost, (1.4% N, 1% P2O5, 1.4% K_20).

Night Soil

- Night soil is human excreta, both solid and liquid, it contain 5.5% N, 4.0% P_2O_5, 2.0% K_2O.

Green Manuring

- Green manuring is the practice of applying as manure large quantities of green material such as leaves and twigs of plants.
- The growing of a crop or applying of leaves and twigs of plant and incorporating it with the soil is called green manuring.

- Green manure crops : Sunnhemp is the green manure crop, it is well suited to almost all parts of country.
- Dhaincha does well in alkaline and water logged soils.
- Lentil is used as a green manure crops in paddy in Kashmir.
- Green Leaf Manuring : When green leafy material is collected form other places and ploughed in the field it is known as green leaf manuring.
- Such crops is *Karanji, Ipomoea* etc.
- Sunnhemp crop is ready for turning in at the age of 7 to 8 weeks whereas dhaincha crop is ready for incorporation when 5-6 weeks old.

Green Manuring : Merits

1. Green manuring adds organic matter and nitrogen to the soil.
2. Green manuring checks weed growth.
3. Green manuring crop aid in the reclamation of saline and alkaline soils.
4. Green manuring increases the availability of plant nutrients.

Relevance and Preparation of Vermicompost

- Vermicompost can be prepared in the field pit. In the lack of space in the garden, it can be started in containers like, earthen pot, metal container, plastic tub, domestic waste and crop residues are to be stored in pit/container. Cow dung and-red earthworms are added into the collected waste.
- The best pH 6.5 to 7.5 for vermicompost prepare.
- Approximately 3 months required for the making of vermicompost.

Average Composition in Diverse Organic Manures

Organic manures	*Nitrogen*	P_2O_5	K_2O
F.Y. M.	0.50	0.25	0.60
Compost	0.89	0.51	3.05
Night soil	5.50	4.0	2.0
Sewage and Sludge	3.0-6.0	2.0	1.0
Poultry manure	3.03	2.63	1.40
Bird guano	7.0-8.0	11.0-14.0	2.0-3.0
Blood meal	13.0-20.0	—	—
Raw bone meal	3.0-4.0	20.0-25.0	—
Fish manure	4.0-10.0	3.0-9.0	0.3-1.5

Classification of Rhizobium species with their respective host crops

Rhizobium species	*Crop inculation group*
R. japonicum	Soyabean group
R. leguminosarum	Pea group
R. lupini	Lupine group
R. meliloty	Alfalfa group
R. phaseoli	Bean group
R. trifoli	Clover group
R. varioue species	Cowpea group

Various Fertilizers and their Role

- Fertilizers are inorganic substances that contain one or more ingredients of plant food in large proportions.
- Fertilizers which contain only one plant nutrient, it are

called a straight fertilizers. Such fertilizers is called nitrogenous, or phosphatic or potassic e.g., Urea.

- Fertilizers which contain two or more nutrients, in one compound is known as complex or compound fertilizers.
- Complex fertilizers are granular and easy to apply. For example D.A.P., ammonium phosphate sulphate.
- Some fertilizers, such as ammonium nitrate are very hygroscopic in nature and rapidly absorb moisture.

Categorising Nitrogenous Fertilizers

- The nitrogenous fertilizers are broadly classified into 4 groups

Nitrate Fertilizers : Nitrate group of fertilizers are soluble in water and hygroscopic nature. These are alkaline in nature and contains less percentage of nitrogen than other groups.

- The nitrate is not absorbed by the soil. They are applied in small doses.
- Nitrate fertilizers like sodium nitrate (16% N), calcium nitrate (15.0% N), potassium nitrate (13% N).

Ammonium Fertilizers : This group of fertilizers are widely used, particularly ammonium sulphate. It is soluble in water but not hygroscopic.

- They are acidic in nature.
- Rice is the merely crop which can utilize the ammoniacal nitrogen.
- This group of fertilizers may be used in basal application and top dressing.

Nitrate and Ammonium Fertilizers : These group of fertilizers are soluble in water but moderately hygroscopic like-ammonium nitrate, ammonium sulphate, etc. in this group both ammonium and nitrate are present.

- It is acidic in nature and used in top dressing and basal dressing.

Amide Fertilizers : It is soluble in water and hygroscopic in nature.

- These fertilizers are converted to ammonium carbonate and then to nitrates due to action of micro-organism.
- The conversion of amides into ammoniacal and nitrate form takes about 6-7 days.
- Amide fertilizers like-urea, calcium cynamid.
- Urea is a organic fertilizer.

Nitrogen Loss in the Field

1. Volatilization loss : In this process nitrogen is lost in the gaseous form.
2. Denitrification loss : The biochemical reduction of nitrate-nitrogen to gaseous compound by micro-organism is called denitrification.
    - It process mostly in water flooded area due to poor drainage by micro-organism.
3. Leaching loss : Urea $[CO(NH_2)_2]$ contain 46% nitrogen.

Fertilizers of Phosphatic Nature

1. Single super phosphate $[Ca(H_2PO_4)_2]$ contain 16-20% P_2O_5.
2. Triple super phosphate $[3Ca(H_2PO_4)_2]$ contain 40-48% P_2O_5.
3. Diammonium phosphate $(NH_4)_2HPO_4$ contain 54% P_2O_5 and 21% nitrogen. It is acidic in nature.
4. Rock phosphate $[Ca_3(PO_4)_2]$ contain P_2O_5 20-30%. It is insoluble fertilizers.
    - Phosphorus fertilizers are classified into three groups: (a) Water soluble (b) Citrate soluble (c) Insoluble
    - Super phosphate : The phosphorus in bones and rock

phosphates is in the form of tricalcium phosphate and therefore, does not become readily available to plants.

- The insoluble phosphate in rock phosphate is converted into a soluble form by treating with sulphuric acid and the resulting product is called super phosphate (ordinary).
- Basic Slag : It is much effective in high rainfall area and that are neutral to acidic in soil reaction. It contain 3-5% P_2O_5. It is alkaline in nature.
- Rock Phosphate: It should be used in heavy application with organic matter for good result.

Potassium based Fertilizers

- Potassium fertilizers are water soluble but not hygroscopic in nature. There are no leachy loss.
- Potassium Chloride : Its use is found effective in acidic and heavy soils.
- Chemical form is KCI.
- Potassium chloride contains 60% K_2O.
- Potassium Sulphate : This fertilizer is most effective on light and medium soils with high pH. It contains 50% K_20.
- The chemical form of potassium sulphate is K_2SO_4.
- Potassium Nitrate (KNO_3) : It is excellent source of K and nitrogen mainly used for fruit trees and vegetables, tobacco etc.
- It contains 44% K_2O and 13% nitrogen.
- Gypsum ($CaSO_4.2H_2O$) contains 29.2% calcium and 18% sulphur.
- Iron pyrite (FeS_2) contains 22-24% sulphur.
- Borax ($Na_2B_4O_7.10H_2O$) contains 10.6% boron.

- Calcium ammonium nitrate (CAN) also called Kisan khad. It contains 25-28% nitrogen and neutral in nature. It is most suitable for vegetables.
- N, P and K also known as complete fertilizer.
- Nitrogen inhibitors is oxamide (31% N), thiourea (36.8%N), dicynamide (42%N), nitropyrin, neem cake.

Soil Fertility and Productivity

1. Soil productivity means the crop producing capacity of a soil which is measured in terms of yields.
2. Soil fertility means the capacity of soil to supply plant nutrient required by a healthy crop.

Different Nutrients

- **Nutrient Determination and Analysing Process.**

Nutrient Determination	*Analysing Process*
1. Total nitrogen in soil	Kzeldahl process
2. Available nitrogen	Alkaline permanganate method
3. Phosphorus in neutral and alkaline condition	Olsen's method
4. K^+ and sodium	Flame photometer
5. Organic carbon	Morgan's method and $KMnO_4$ process
6. Available potash	Chromic acid digestion method

Classification of Essential Nutrients

1. Basic Nutrients : C, H, O is the basic nutrients because its constitute about 96% of the total dry matter of plants.

2. Major Nutrients : Which elements required in larger amount is called macro or major nutrients. These are N, P, K Ca, Mg and S.
   - Ca, Mg and S is secondary nutrients.
3. Micro Nutrients : Such elements are required in small quantities (0.01-100 ppm)
   - Micro nutrients are also called trace elements oligoelements.
   - Micro nutrients are Mn, Cu, Mo, B, CI, Co, Fe and Zn.
   - According to mobility in plants, nutrients classified are:

     (a) Highly mobile — N, P and K

     (b) Moderately mobile — Zn

     (c) Less mobile — S, Fe, Cu, Mn, Mo and Cl

     (d) Immobile — Ca and B
   - Another classification based on functions of elements:

     (a) Energy exchangers — H, O

     (b) Energy stores — C, N, P and S

     (c) Translocation regulators — K, Na, Ca and Mg

     (d) Oxidation reduction — Fe, Mn, Mo, Cu, B and Zn
   - Mycorrhiza : The beneficial association between a plant root and its endophytic fungal associate is known as Vascular Arbuscular Mycorrhiza (VAM), which helps in increasing phosphorus availability.

IAS 2002 U.P. PCS (Pre 2001).

Glossary

- Available Water: The portion of water in a soil that can be readily absorbed by plant roots. The amount of water released between the field capacity and the permanent wilting point.
- Available Nutrient : That portion of any element or compound in the soil that can be readily absorbed and assimilated by growing plants. ("Available" should not be confused with "exchangeable").
- Autotroph : An organism capable of utilizing carbon dioxide or carbonates as the sole source of carbon and obtaining energy for life processes from the oxidation of inorganic elements or compounds such as iron, sulphur, hydrogen, ammonium and nitrites or from radiant energy. Contrast with heterotroph.
- Autochthonous Organisms : These micro-organisms thought to subsist on the more resistant soil organic matter and little affected by the addition of fresh organic materials. Contrast with zymogenous organisms.
- Atterberg Limits : Water contents of fine-organised soils at different states of consistency.
- Liquid Limit (L.L.) : The water content corresponding to the arbitrary unit between the liquid and plastic states of consistency of a soil.
- Plastic Limit (PL.) : The water content corresponding to an arbitrary limit between the plastic and semi solid states of consistency of a soil.
- Aspect (of slopes) : The direction (e.g., south or north) that a slope faces with respect to the sun.
- Aridisols : An order in soil Taxonomy. Soils of dry climates. They have pedogenic horizons, low in organic matter, that are never moist for as long as three

consecutive months. They have an ochric epipedon and one or more of the following diagnostic horizons: argillic, nitric, cambic, calcic, petrocalcic, gypsic, petrogypsic, salic or a duripan.

- Arid Climate : Climate in regions that lack sufficient moisture for crop production without irrigation. In cool regions annual precipitation is usually less than 25 cm. It may be as high as 50 cm. in tropical regions. Natural vegetation is desert shrubs.
- Argillic Horizon : A diagnostic subsurface horizon characterized by the illuvial accumulation of layer-lattice silicate clays.
- Argillan : A thin coating of well-oriented clay particles on the surface of a soil, aggregate, particle or pore. A clay film.
- Arbuscule : Specialized branched structure formed with in a root cortical cell by endotrophic mycorrhizal fungi.
- Arbuscular Mycorrhiza : A common endomycorrhizal association produced by phycomycetous fungi and characterized by the development, with in root cells, of small structures known as arbuscules. Some also form between root cells, storage organs known as vesicles. Host range includes many agricultural and horticultural crops. Formerly called Vesicular Arbuscular Mycorrhiza (VAM).
- Aquifer: A saturated, permeable layer of sediment or rock that can transmit significant quantities of water under normal pressure conditions.
- Aguiclude : A saturated body of rock or sediment that is incapable of transmitting significant quantities of water under ordinary water pressures.
- Aquic Conditions : Continuous or periodic saturation

(with water) and reduction. Commonly indicated by redoximorphic features.

- Apatite : A naturally occurring complex calcium phosphate that is the original source of most of the phosphate fertilizers. Formula such as $[3Ca_3(PO_4)_2]. CaF_2$ illustrate the complex compounds that make up apatite.
- Ap : The surface layer of a soil disturbed by cultivation or pasturing.
- Antibiotic : A substance produced by one species of organism that in low concentrations will kill or inhibit growth of certain other organisms.
- Anthropic Epipedon : A diagnostic surface horizon of mineral soil that has the same requirements as the mollic epipedon but that has more than 250 mg/kg of P_2O_5 soluble in 1% citric acid, or is dry more than 10 months (cumulative) during the period when not irrigated. The anthropic epipedon forms under long continued cultivation and fertilization.
- Anion Exchange Capacity : The sum total of exchangeable anions that a soil can absorb. Expressed as centimoles of charge per kilogram ($Cmol_c/Kg$) of soil (or of other adsorbing material, such as clay).
- Anion Exchange : Exchange of anions in the soil solution for anions adsorbed on the surface of clay and humus particles.
- Anion : Negatively charged ion; during electrolysis it is attracted to the positively charged anode.
- Angle of Repose : The maximum slope steepness at which loose, cohesionless material will come to rest.
- Andisols : An order in soil Taxonomy. Soils developed from volcanic ejects. The colloidal fraction is dominated by allophane and/or Al-humus compounds.

- Andic Properties : Soil properties related to volcanic origin of materials, including high organic carbon content, low bulk density, high phosphate retention and extractable iron and aluminium.
- Anaerobic: (a) Without molecular oxygen, (b) Living or functioning in the absence of air or free oxygen.
- Amorphous Material : Non-crystalline constituents of soils.
- Ammonium Fixation : The entrapment of ammonium ions by the mineral or organic fractions of the soil in forms that are insoluble in water and are at least temporarily nonexchangeable.
- Ammonification : The biochemical process where by ammoniacal nitrogen is released from nitrogen-containing organic compounds.
- Amino Acids : Nitrogen-containing organic acids that couple together to form protein. Each acid molecule contains one or more amino groups ($—NH_2$) and at least one carboxyl group (-COOH). In addition, some amino acids contain suffer.
- Amendment Soil : Any substance other than fertilizers, such as lime, sulphur, gypsum and sawdust, used to alter the chemical or physical properties of a soil, generally to make it more productive.
- Aluminosilicates : Compounds containing aluminium, silicon and oxygen as main constituents. An example is microcline, $KAISi_3O_8$.
- Alpha Particle : A positively charged particle (consisting) of two protons and two neutrons that is emitted by certain radioactive compounds.
- Alluvium : A general term for all detrital material deposited or in transit by steams, including gravel,

sand, silt, clay and all variations and mixtures of these. Unless otherwise noted, alluvium is unconsolidated.

- Alluvial Fan : Fan-shaped alluvium deposited at the mouth of a canyon or ravine where debris-laden waters fan out, slow down, and deposit their burden.
- Allophane : A poorly defined aluminosilicate mineral whose structural frame work consists of short runs of three-dimensional crystals interspersed with amorphous noncrystalline materials. Along with its more weathered companion, it is prevalent in volcanic ash materials.
- Allelopathy : The process by which one plant may affect other plants by biologically active chemicals introduced into the soil either directly by leaching or exudation from the source plant, or as a result of the decay of the plant residues. The effects, though usually negative, many also be positive.
- Allelochemical : An organic chemical by which one plant can influence another.
- Alkaline Soil : Any soil that has pH > 7. Usually applied to the surface layer or root zone but may be used to characterize any horizon or a simple there of.
- Algal Bloom: A population explosion of algae in surface waters, such as lakes and steams often resulting in high turbidity and green or red-coloured water, commonly stimulated by nutrient enrichment with phosphorus and nitrogen.
- Alfisols : An order in soil Taxonomy. Soils with gray to brown surface horizons, medium to high supply of bases, and B horizons of illuvial clay accumulation. These soils form mostly under forest or savanna vegemoisture deficit.
- Albic Horizon : A diagnostic sub-surface horizon from which clay and free iron oxides have been segregated

to the extent that the colour of the horizon is determined primarily by the colour of the primary sand and silt practicles rather than by coatings on these particles.

- Air Porosity : The proportion of the bulk volume of soil that is filled with the air at any given time or under a given condition, such as a specified moisture potential; usually the large pores.
- Air Dry : (a) The state of dryness (of a soil) at equilibrium with the moisture content in the surrounding atmosphere. The actual moisture content will depend upon the relative humidity and the temperature of the surrounding atmosphere. (b) To allow to reach equilibrium in moisture content with the surrounding atmosphere.
- Agronomy : A specialization of agriculture concerned with the theory and practice of field-crop production and soil management. The scientific management of land.
- Agric Horizon : A diagnostic subsurface horizon in which clay, silt, and humus derived from an overlying cultivated and fertilized layer have accumulated. Worm holes and illuvial clay, silt and humus occupy at least 5% of the horizon by volume.
- Aggregate (soil) : Many soil particles held in a single mass or cluster, such as a cold, crumb, block or prism.
- Aerosolic Dust : A type of eolian material that is very fine. (about 1 to 10 (μm) and may remain suspended in the air over distances of thousands of kilometers. Finer than most loess.
- Aerobic : (a) Having molecular oxygen as a part of the environment. (b) Growing only in the presence of molecular oxygen, as aerobic organisms. (c) Occurring only in the presence of molecular oxygen (said of certain

chemical or biochemical processes, such as aerobic decomposition).

- Aeration Soil : The process by which air in the soil is replaced by air from atmosphere. In a well-aerated soil, the soil air is similar in composition to the atmosphere above the soil. Poorly aerated soils usually contain more carbon dioxide and correspondingly less oxygen than the atmosphere above the soil.
- Aerate : To impregnate with gas, usually air.
- Adsorption Complex : The group of organic and inorganic substances in soil capable of adsorbing ions and molecules.
- Adsorption : The attraction of ions or compound to the surface of a solid. Soil colloids adsorb large amounts of ions and water.
- Adhesion : Molecular attraction that holds the surfaces of two substances (e.g., water and sand particles) in contact.
- Active Organic Matter : A portion of the soil organic matter that is relatively easily metabolized by micro-organisms and cycles with a half-life in the soil of a few days to a few years.
- Activated Sludge : Sludge that has been aerated and subjected to bacterial action.
- Actinomycetes : A group of organisms intermediate between the bacteria and the true fungi that usually produce a characteristic branched mycelium. Includes many, but not all, organisms belonging to the order of Actinomycetales.
- Acidity Total : The total acidity in a soil. It is approximated by the sum of the salt-replaceable acidity plus the residual acidity.

- Acidity Salt Replaceable : Exchangeable hydrogen and aluminium that can be replaced from an acid soil by an unbuffered salt solution such as KCI or NaCl.
- Acidity Residual : Soil acidity that can be neutralized by lime or other alkaline materials but can not be replaced by an unbuffered salt solution.
- Acidity, Active : The activity of hydrogen ions in the aqueous phase of a soil. It is measured and expressed as a pH value.
- Acid Sulphate Soils : Soils that are potentially extremely acid ($pH < 3.5$) because of the presence of large amounts of reduced forms of sulphur that are oxidized to sulphuric acid if the soils are exposed to oxygen when they are drained or excavated. A sulphuric horizon containing the yellow mineral jarosite is often present.
- Acid Soil : A soil with pH value < 7.0. Usually applied to surface layer or root zone, but may be used to characterize any horizon.
- Acid Saturation : The proportion or percentage of a cation-exchange site occupied by acid cations.
- Acid Rain : Atmospheric precipitation with pH values less than about 5.6, the acidity being due to inorganic acids such as nitric and sulphuric that are formed when oxides of nitrogen and sulphur are emitted into the atmosphere.
- Acid Cations : Cations, principally $Al^{3+}Fe^{3+}$ and H^+ that contributes to H^+ ion activity either directly or through hydrolysis reactions with water.
- Accelerated Erosion : Erosion much more rapid than normal, natural, geological erosion; primarily as a result of the activities of humans or, in some cases, of animals.
- Abiotic : Non-living basic elements of the environment such as rain fall, temperature, wind and minerals.

- A-Horizon : The surface horizon of a mineral soil having maximum organic matter accumulation, maximum biological activity and/or eluviation of materials such as iron and aluminium oxides and silicate clays.
- Buried Soil : Soil covered by an alluvial, loessal or other deposit usually to a depth greater than the thickness of the solum.
- Bulk Density, Soil : The mass of dry soil per unit of bulk volume including the air space. The bulk volume is determined before drying to constant weight at 105°C.
- Bulk Blending : Mixing dry individual granulated fertilizer materials to form a mixed fertilizer that is applied promptly to the soil.
- Buffering Capacity : The ability of a soil to resist changes in pH. Commonly determined by presence of clay, humus and other colloidal materials.
- Brown Fields : Abandoned, idled or under used industrial and commercial facilities where expansion or redevelopment is complicated by real or perceived environmental contamination.
- Broadcast : Scatter seed or fertilizer on the surface of soil.
- Broad-base Terrace : A low embankment with such gentle slopes that it can be formed, constructed across sloping fields to reduce erosion and run off.
- Blocky Soil Structure : Soil aggregates with block like shapes; common in 13-horizons of soils in humid regions.
- Biostimulation : The clean up of contaminated soils through the manipulation of nutrients or other soil environmental factors to enhance the activity of naturally occurring soil micro-organisms. A form of bioremediation.

- Biosolids : Sewage sludge that meets certain regulatory standards, making it suitable for land application.
- Biosequence : A group of related soils that differ,: one from the other, primarily because of differences in kinds and numbers of plants and soil organisms as a soil-forming factor.
- Bioremediation : The decontamination or restoration of polluted or degraded soils by means of enhancing the chemical degradation or other activities of soil organisms.
- Biopores : Soil pores, usually of relatively large diameter, created by plant roots, earthworms or other soil organisms.
- Biomass : The total mass of living material of a specified type (e.g., microbial biomass) in a given environment (e.g., in a cubic meter of soil).
- Biological Nitrogen Fixation : Occurs at ordinary temperatures and pressures. It is commonly carried out by certain bacteria, algae and Actinomycetes which may or may not be associated with higher plants.
- Biodegradable : Subject to degradation by biochemical processes.
- Beta Particle : A high-speed electron emitted in radioactive decay.
- Bioaugmentation : The clean up of contaminated soils by adding exotic micro-organisms that are especially efficient at breaking down an organic contaminant. A form of bioremediation.
- Bio-accumulation : A building within an organisms of specific compounds due to biological processes commonly applied to heavy metals, pesticides or metabolites.

- Bench Terrance : An embankment constructed across sloping fields with a steep drop on the down slope side.
- Bed Rock: The solid rock underlying soils and the regolith in depths ranging from zero (where exposed by erosion) to several hundred feet.
- Bedding : (Engineering) Arranging the surface of fields by plowing and grading into a series of elevated beds separated by shallow depression or ditches for drainage.
- Base Saturation Percentage : The extent to which the adsorption complex of a soil is saturated with exchangeable cations other than hydrogen and aluminium. It is expressed as a percentage of the total cation exchange capacity.
- Base-forming Cations (Obsolete) : Those cations that form strong (strongly dissociated) bases by reaction with hydroxy; e.g., K' forms potassium hydroxide (K^{+}+$OH^{)}$.
- Bar : A unit of pressure equal to 1 million dyness per square centimeter (10^{6} dynes/cm^{2}). It approximates the pressure of a standard atmosphere.
- B-Horizon : A soil horizon, usually beneath the A-horizon that is characterized by one or more of the following: (a) A concentration of silicate clays, iron and aluminium oxides and humus alone or in combinations; (b) A blocky or prismatic structure. (c) Coatings of iron and aluminium oxides that give darker, stronger or redder colour.
- Bulk Blended Fertilizers : Solid fertilizer materials blended together in small blending plants. Delivered to the farm in bulk and usually special applicator.
- C-Horizon : A mineral horizon generally beneath the solum that is relatively unaffected by biological activity and pedogenesis and is lacking properties diagnostic

of an A or B-horizon it may or may not be like the material form which the A and B have formed.

- Calcaxeous Soil : Soil containing sufficient calcium. Carbonate (often with magnesium carbonate) to efferuese visibly when treated with cold 0.1 N hydrochloric acid.
- Calcic Horizon : A diagnostic subsurface horizon of secondary carbonate enrichment that is more then 15 cm. thick has a calcium carbonate equivalent of more than 15% and has at least 5% more calcium carbonate equivalent then the underlying C-horizon.
- Caliche : A layer near the surface more or less commented by secondary carbonates of calcium or magnesium. Precipitated from the soil solution it may occur as a soft then soil horizon as a hard thick bed just beneath the solum or as a surface layer exposed by erosion.
- Cambic Horizon : A diagnostic subsurface horizon that has a texture of loamy very fine sand or finer contains some weatherable minerals and is characterized by the alteration or removal of mineral material the cambic horizon lacks cementation or induration and has too few evidences of illuviation to meet the requirements of the argillic or spodic horizon.
- Capillary Water : The water held in the capillary or small pores of a soil usually with a tension 60 cm. of water.
- Carbon Cycle : The sequence of transformations whereby carbon dioxide is fixed in living organisms by photosynthesis or by chemosynthesis liberated by respiration and by the death and decomposition of the fixing. Organism used by heterophic species and ultimately returned to its original state.

- Cost Earthworm : Rounded water-stable aggregates of soil that have passed through the gut of an earthworm.
- Cat Clays : Wet clay soil high in reduced forms of sulphur that upon being drained become extremely acid because of the oxidation of the sulphur compounds and the formation of sulphuric acid usually found in tidal marshes.
- Catena : A sequence of soil of about the same age derived from similar parent material and occurring under similar climatic conditions but having different characteristics because of variation in relief and in drainage.
- Cation : A positively charged ion during electrolysis it is attracted to the negatively charged cathode.
- Cation Exchange: The interchange between a cation in solution and another cation on the surface of any surface active material such as clay or organic matter.
- Chert : A structureless form of silica closely related to flint that breaks into angular fragments.
- Cation Exchange Capacity : The sum total of exchange able cations that a soil can absorb sometimes called total exchange capacity base exchange capacity or cation adsorption capacity expressed in contimoles of charge per kilogram (C mol_c/Kg) of soil (or of other absorbing material such as clay).
- Cemented : Indurated having a hard brittle consistency because the particles are held together by cementing substances such as humus calcium carbonate or the oxides of silicon iron and aluminium.
- Chelate (Greek Claw) : A type of chemical compound in which a metallic ion is firmly combined with an organic molecule by means of multiple chemical bonds.

- Chisel Subsoil : A tillage implement with one or more cultivator type feet which are attached strong knife like units used to shatter or loosen hard compact layers usually in the subsoil to depths below normal plow depth.
- Chlorite : A 2 : 1 : 1-type layer-structured silicate, mineral having 2 : 1 layers *alternating* with a magnesium, dominated octahedral sheet.
- Clay : (a) A soil separate consisting of particles < 0.002 mm in equivalent diameter (b) A soil textural class *containing* > 40% clay < 45% sand and < 40% silt.
- Clay Mineral : Naturally occurring inorganic material (usually crystalline) found in soil and other earthy deposits the particles being of clay size that is < 0.002 mm. in diameter.
- Clod : A compact cohexent mass of soil produced artificially usually be such human activities as plowing and digging especially when these operations are performed on soils that are either too wet or too dry for normal tillage operations.
- Coarse Texture : The texture exhibited by sands loamy sands and sandy loam (except very fine sandy loam).
- Cobblestone : Rounded or partially rounded rock or mineral fragments 7.5 of 25 cm. (3 to 10 in) in diameter.
- Collapsible Soil : Certain soil that many undergo a sudden loss in strength when wetted.
- Colloid Soil (Greek glue like) : Organic and inorganic matter with very small particle size and a correspondingly layer surface area per unit of mass.
- Compost : Organic residues or a mixture of organic residues and soil that have been piled moistened and allowed to undergo biological decomposition mineral

fertilizers are sometimes added often called artificial manure or synthetic manure if produced primarily from plant residues.

- Concretion : A local concentration of a chemical compound such as calcium carbonate or iron oxide in the form of grains or nodules of varying size shape hardness and colour.
- Conduction : The transfer of heat by physical contact between two or more objects.
- Consistency : The interaction of adhesive and cohesive forces within a soil at various moisture contents as expressed by the relative case with which the soil can be deformed or ruptured.
- Consistence : The combination of properties of soil material that determine its resistance to crushing and its ability to be molded or changed in shape such terms as loose friable firm soft plastic and sticky describe soil consistence.
- Consolidation Test : A laboratory test in which a soil mass is laterally confined within a ring and is compressed with a known force between two porous plates.
- Constant Charge : The net surface charge of mineral particles the magnitude of which depends only on the chemical and structural composition of the mineral. The charge arises from isomorphous substitution and is not affected by soil pH.
- Consumptive Use : The water used by plants in transpiration and growth plus water vapour loss from adjacent soil or snow or from intercepted. Precipitation in any specified time usually expressed as equivalent depth of free water per unit of time.

- Contour : An imaginary line connecting points of equal elevation on the surface of the soil. A contour terrae is laid out on a sloping soil at rigles to the direction of the slope and nearly level throughout its course.
- Connection : The transfer of heat through a gas or solution because of molecular movement.
- Creep : Slow mass movement of soil and soil material down relatively steep slopes primarily under the influence of gravity but facilitated by saturation with water and by alternate freezing and thawing.
- Crushing Strength : The force required to crush a mass of dry soil or conversely the resistance of the dry soil mass to crushing expressed in units of force per unit area (Pressure).
- Crust : A surface layer on soil ranging in thickness from a few millimeters to perhaps as much as 3 cm. that is much more compact hard and brittle when dry than the material immediately beneath it.
- Cryoturbation : Physical disruption and displacement of soil material within the profile by the forces of freezing and thawing sometimes called frost churning. It results in irregular broken horizons involution oriented rock fragments and accumulation of organic matter on the permafrost table.
- Crystal Structure : The orderly arrangement of atoms in a crystalline material.
- Crystal: A homogeneous inorganic substance of definite chemical composition bounded by plane surfaces that from definite angles with each other thus giving the substance a regular geometrical form.
- Decomposition : Chemical breakdown of a compound (a mineral or organic compound) into simpler

compounds often accomplished with the aid of micro-organisms.

- Delta : An eluvial deposit formed where a stream or river drops its sediment load upon entering a quieter body of water.
- Denitrification : The biochemical reduction of nitrate or nitrite to gaseous nitrogen either as molecular nitrogen or as an oxide of nitrogen.
- Desalinization : Removal of salts from saline soil usually by leaching.
- Desert Crust : A hard layer containing calcium carbonate gypsum or other binding material exposed at the surface in desert regions.
- Desert Pavement : A natural residual concentration of closely packed pebbles boulders and other rock fragments on a desert surface where wind and water action has removed all smaller particles.
- Diagnostic Horizons : Horizons having specific soil characteristics that are indicative of certain classes of soil horizons that occur at the soil surface are called epipedons those below the surface diagnostic sub-surface horizons.
- Dioctahedral Sheet : An octahedral sheet of silicate clays in which the sites for the six co-ordinated metallic atoms are mostly filled with trivalent atoms.
- Drainage Soil : The frequency and duration of periods when the soil is free from saturation with water.
- Disintegration : Physical or mechanical break up or separation of a substance into its component parts rock breaking into its mineral components.
- Disperse : (a) To break up compound particles such as aggregates into the individual component particles. (b)

To distribute or suspend fine particles such as clay in or throughout a dispersion medium such as water.

- Duripan : A diagnostic subsurface horizon that is cemented by silica to the point that air dry fragments will not slake in water or HCl hardpan.
- E-horizon : Horizon characterized by maximum illuviation (washing out) of silicate clays and iron and aluminium oxides commonly occurs above the B horizon and below the A horizon.
- Earthworms : Animals of the lumbricid family that burrow into and live in the soil they mix plant resides into the soil and improve soil aeration
- Edaphology : The science that deals with the influence of soils on living things particularly plants including human use of land for plant growth.
- Electrical Conductivity : The capacity of a substance to conduct or transmit electrical current in soils or water measured in siemens/meter and related to dissolved solutes.
- Eluviation : The removal of soil material in suspension (or in solution) from a layer or layers of a soil (usually the loss of material in solution is described by the term leaching).
- Entisols : An order in soil Taxonomy soils have no diagnostic pedogenic horizons they may be found in virtually any climate on very recent geomorphic surfaces.
- Eolian Soil Material : Soil material accumulated through wind action the most extensive areas in the united states are silty deposits but large areas of sandy deposits also occur.
- Epipedon : A diagnostic surface horizon that includes

the upper part of the soil that is darkened by organic matter or the upper alluvial horizons or both.

- Equilibrium Phosphorus Concentration : The concentration of phosphorus in a solution in equilibrium with a soil the EPCo being the concentration of phosphorus achieved by desorption of phosphorus from a soil to phosphorus free distilled water.
- Erosion : (a) The wearing away of the land surface by running water wind ice or other geological agents including such processes as gravitational creep. (b) Detachment and movement of soil or rock by water wind ice or gravity.
- Exchange Capacity : The total ionic charge of the adsorption complex active in the adsorption of ions see also anion exchange capacity cation exchange capacity.
- Exchangeable Ions : Positively or negatively charged atoms or groups of atoms that are held on or near the surface of a solid particle by attraction to charges of the opposite sign and which may be replaced by other like charged ions in the soil solution.
- Exchangeable Sodium Percentage : The extent to which the adsorption complex of a soil is occupied by sodium it is expressed as follows

$$ESP = \frac{\text{Exchangeable Sodium (Cmol}_c\text{/kg) soil}}{\text{Cation Exchange capacity (C mol}_c\text{/kg) soil}} \times 100$$

- Exfoliation : Peeling away of layers of a rock from the surface inward usually as the result of expansion and contraction that accompany changes in temperature.
- Expansive Soil : Soil that undergoes significant volume change upon wetting and drying usually because of a high content of swelling type clay minerals.
- External Surface : The area of surface exposed, on the top bottom and sides of a clay crystal or micelle.

- Family Soil : In soil Taxonomy one of the categories intermediate between the great group and the soil series families are defined largely on the basis of physical and mineralogical properties of importance to plant growth.
- Ferrihydrite : $Fe_2HO_8.4H_2O$. A dark reddish brown poorly crystalline. Iron oxide that forms in wet soils.
- Fertiligation : The application of fertilizers in irrigation waters commonly through sprinkler systems.
- Fertilizer : Any organic or inorganic material of natural or synthetic origin added to a soil to supply certain elements essential to the growth of plants the major types of fertilizers include.
- Fertilizer Requirement : The quantity of certain plant nutrient elements needed in addition to the amount supplied by the soil to increase. Plant growth to a designated optimum.
- Fertility Soil : The quality of a soil that enables it to provide essential chemical elements in quantities and proportions for the growth of specified plants.
- Fine Texture : Consisting of or containing large quantities of the fine fractions particularly of silt and clay (includes clay loam, sandy clay loam, silty clay loam, sandy clay silty clay and clay, textural classes).
- Fine Grained mica : A silicate clay having a 2 : 1 type lattice structure with much of the silicon in the tetrahedral sheet having been replaced by aluminium and with considerable interlayer potassium which binds the layers together prevents interlayer expansion and swelling and limits interlayer cation exchange capacity.
- Fixation : (a) For other then elemental nitrogen : The process or processes in a soil by which certain chemical elements essential for plant growth are converted from

a soluble or exchangeable form to a much less soluble or to a non-exchangeable form for example potassium ammonium and phosphate fixation. (b) For elemental nitrogen process by which gaseous elemental nitrogen is chemically combined with hydrogen to for ammonia.

- Flagstone : A relatively thin rock or mineral fragment 15 to 38 cm. in length commonly composed of shale slate limestone or sandstone.
- Flocculate : To aggregate or clump together individual tiny soil particles especially five clay into small clumps or flocculus opposite of deflocculate or disperse.
- Fluorapatite : A member of the apatite group of minerals containing fluorine most common mineral in phosphate rock.
- Fluvial Deposits : Deposits of parent materials laid down by rivers or streams.
- Fragipan : Dense and brittle pan or sub-surface layer in soil that owes its hardness mainly to extreme density or compactness rather than high clay content or cementation removed fragments are friable but the material in place is so dense that roots penetrate and water moves through it very slowly.
- Friable : A soil consistency term pertaining to soils that crumble with ease.
- Frigid : A soil temperature class with mean annual temperature below 8°C.
- Fulvic Acid : A term of varied usage but usually referring to the mixture of organic substances remaining in solution upon acidification of a dilute alkali extract from the soil.
- Functional Group : An atom or group of atoms attached to a large molecule each functional group (e.g., OH—CH_3—COOH) has a characteristic chemical reactivity.

- Furrow Slice : The uppermost layer of an arable soil to the depth of primary village for layer of soil sliced away from the rest of the profile and inverted by a moldboard plow.
- Granulated Fertilizers : Solid fertilizer materials blended together in small blending plants size which facilitate ease of handing the materials and reduce undesirable dusts.
- Gelisols : An order in soil Taxonomy soil that have permafrost within the upper 1 m. or upper 2 m. if cry oturbation is also present they may have an ochric histic mollic or other epipedon.
- Genetic Horizon : Soil layers that resulted from soil forming processes as apposed to sedimentation or other geologic processes.
- Genesis Soil : The mode of origin of the soil with special reference to the processes responsible for the development of the solum or true soil from the unconsolidated parent material.
- Geological Erosion : Wearing away of the earth & surface by water ice or other natural agents. Under natural environmental conditions of climate vegetation and so on undisturbed by man synonymous with natural erosion.
- Gibbsite $Al(OH)_3$: An aluminium trihydroxide mineral most common in highly weathered soil such as oxisols.
- Glaciofluvial Deposits : Material moved by glaciers and subsequently sorted and deposited by streams flowing from the melting ice the deposits are stratified and may occur in the form of outwash plains deltas kames eskers and kame terraces.
- Gleyed : A soil condition resulting from prolonged saturation with water and reducing conditions that

manifest themselves in greenish or bluish colours throughout the soil mass in mottles.

- Granular COOH : A yellow brown iron oxide mineral that accounts for the brown colour in many soil.
- Granular Structure : Soil structure in which the individual grains are grouped into spherical aggregates with indistinct sides highly porous granules are commonly called crumbs. A well granulated soil has the best structure for most ordinary crop plants.
- Granulation : The process of producing granular materials commonly used to refer to the formation of soil structural granules but also used to refer to the processing of powdery fertilizer materials into granules.
- Gravitational Potential : That portion of the total soil water potential due to differences in elevation of the reference pool of pure water and that of the soil water since the soil water elevation is usually chosen to be higher than that of the reference pool and gravitational potential is usually positive.
- Gravitational Water : Water that moves into through or out of the soil under the influence of gravity.
- Green House Effect : The entrapment of heat by upper atmosphere gases such as carbon dioxide water vapour and methane just as glass traps heat for a green house increases in the quantities of these gases in the atmosphere will likely result in global warming that may have serious consequences for human kind.
- Granulated Fertilizers : Fertilizers that are present in the form of rather stable granites of uniform size which facilitate ease of handling the materials and reduce undesirable dusts.
- Ground Water : Sub-surface water in the zone of

saturation that is free to move under the influence of gravity often horizontally to stream channels.

- Gully Erosion : The erosion process whereby water accumulates in narrow channels and over short periods removes the soil from this narrow area to considerable depths ranging from 1 to 2 ft. to as much as 23 to 30 m.
- Gypsic Horizon : A diagnostic sub-surface horizon of secondary calcium sulphate environment that is more than 15 cm. thick.
- Gypsum Requirement : The quantity of gypsum required to reduce the exchangeable sodium percentage in a soil to an acceptable level.
- Hardpan : A hardened soil layer in the lower A or in the B-horizon caused by cementation of soil particles with organic matter or with such materials as silica sesquioxides or calcium carbonate the hardness does not change appreciably with changes in moisture content and pieces of the hard layer do not slake in water.
- Heavy Soil : A soil with a high content of the fine separates particularly clay or one with a high drawbar pull hence, difficult to cultivate.
- Heavy Metals : Those metals that have densities of 5.0 mg/m or greater elements in soil include Cd, Co, Cr, Cu, Fe, Hg, Mn, Mo, Pb and Zn.
- Hematite Fe_2O_3 : A red iron oxide mineral that contributes red colour to many soils.
- Histic Epipedon : A diagnostic surface horizon consisting of a thin layer of organic soil material that is saturated with water at some period of the year unless artificially drained and that is at or near the surface of a mineral soil.
- Histosols : An order in soil Taxonomy soil formed from materials high in organic matter histosols with

essentially no clay must have at least 20% organic matter by weight (about 78% by volume) the minimum organic matter content rises with increasing clay content to 30% (85% by volume) in soil with at least 60% clay.

- Horizon Soil : A layer of soil approximately parallel to the soil surface differing in properties and characteristics from adjacent layers below or above it.
- Humic Acid : A mixture of variable or indefinite composition of dark organic substances precipitated upon acidification of a dilute alkali extract from soil.
- Humic Substances : A series of complex relatively high molecular weight brown to black coloured organic substances that make up 60 to 80% of the soil organic matter and generally quite resistant to ready microbial attack.
- Humification : The processes involved in the decomposition of organic matter and leading to the formation of humus.
- Humin : The fraction of the soil organic matter that is not dissolved upon extraction of the soil with dilute alkali.
- Humus : That more or less stable fraction of the soil organic matter remaining after the major portions of added plant and animal residues. Have decomposed usually it is dark in colour.
- Hydration : Chemical union between on ion or compound and one or more water molecules the reaction being stimulated by the attraction of the ion or compound for either the hydrogen or the unshared electrons of the oxygen in the water.
- Hydraulic Conductivity : An expression of the readiness with which a liquid such as water flows through a solid such as soil in response to a given potential gradient.

- Hydrolysis : A reaction with water that splits the water molecule into H^+ and OH^- ions. Molecules or atoms participating in such reaction are said to hydrolyze.
- Hydronium : A hydrated hydrogen ion (H_3O_+) the form of the hydrogen ion usually found in an aqueous system.
- Hydroxy Apatite : A member of the apatite group of minerals rich in hydroxyl groups. A nearly insoluble calcium phosphate.
- Hygroscopic Coefficient: The amount of moisture in a dry soil when it is in equilibrium with some standard relative humidity near a saturated atmosphere (about 98%) expressed in terms of percentage on the basis of oven day soil.
- Igneous Rock : Rock formed from the cooling and solidification of magma that has not been changed appreciably since its formation.
- Illuvial Horizon : A soil layer or horizon in which material carried from an overlying layer has been precipitated from solution or deposited from suspension the layer of a accumulation.
- Immature Soil : A soil with indistinct or only slightly developed horizons because of the relatively short time it has been subjected to the various soil forming processes. A soil that has not reached equilibrium with its environment.
- Indurated (Soil) : Soil material cemented into the hard mass that will not soften on wetting.
- Infiltration : The downward entry of water into the soil.
- Inorganic Compounds : All chemical compounds in nature except compounds of carbon other than carbon monoxide carbon dioxide and carbonates.
- Infiltration Capacity : A soil characteristic determining

or describing the maximum rate at which can enter the soil under specified conditions including the presence of an excess of water.

- Integrade : A soil that possesses moderately well developed distinguishing characteristics of two or more genetically related great soil groups.
- Internal Surface : The area of surface exposed within a clay crystal or micelle between the individual crystal layers compare with external surface.
- Inter Layer : Materials between layers within a given crystal including cations hydrated cations organic molecules and hydroxide groups or sheets.
- Ions : Atoms groups of atoms or compounds. That are electrically charged as a result of the loss of electrons or the gain of electrons (anions).
- Isotopes : Two or more atoms of the same element that have different atomic masses because of different numbers of neutrons in the nuclear.
- joule : The SI energy unit defined as a force of 1 newton applied over a distance of 1 metre 1 joule = 0.239 calorie.
- Kandic Horizon : A sub-surface diagnostic horizon having a sharp clay increase relative to overlying horizons and having low activity clays.
- Kaolinite : An aluminosilicate mineral of the 1 : 1 crystal lattice group that is consisting of single silicon tetrahedral sheets alternating of with single aluminium octahedral sheets.
- Land : Abroad term embodying the total natural environment of the areas of the earth not covered by water in addition to soil its attributes include other physical conditions such as mineral deposits and water supply location in relation to centers of commerce

population and other land the size of the individual tracts or holdings and existing plant cover works of improvement and the like.

- Land Classification : The arrangement of land units into various categories based upon the properties of the land or its suitability for some particular purpose.
- Land Forming : Shaping the surface of the land by scraping off the high spots and filling in the low spots with precision grading machinery to create a uniform smooth slope often for irrigation purposes also called land smoothing.
- Land Use Planning : The development of plant for the uses of land that over long periods will best serve the general welfare together with the formulation of ways and means for achieving such uses.
- Laterite : An iron rich sub-soil layer found in some highly weathered humid tropical soil that when exposed and allowed to dry becomes very hard and will not soften when rewedded when erosion removes. The overlying layers the laterite is exposed and a virtual pavement results.
- Layer (Clay mineralogy) : A combination in silicate clays of (tetrahedral and octahedral) sheets in a 1 : 1, 2 : 1 or 2 : 1 : 1 combination.
- Leaching : The removal of materials in solution from the soil by percolating water.
- Leaching Requirement : The loathing fraction of irrigation water necessary to keep soil salinity from exceeding a tolerance level of the crop to be grown.
- Liebig and Law : The growth and reproduction of an organism are determined by the nutrient substance (oxygen, carbon dioxide, calcium etc.) that is available

in minimum quantity with respect to organic. Needs the limiting factor.

- Light Soil (Obsolete in scientific use) : A coarse textured soil a soil with a low drawbar pull and hence, easy to cultivate.
- Lignin : The complex organic constituent of woody fibres in plant tissue that along with cellulose cements the cells together and provides strength lignins resist microbial attack and after some modification may become part of the soil organic matter.
- Lime (Agricultural) : In strict chemical terms calcium oxide in practical terms a material containing the carbonates oxides and or hydroxides of calcium and/or magnesium used to neutralize soil acidity.
- Lime Requirement : The mass of agricultural limestone of the equivalent of other specified liming material required to raise the pH of the soil to a desired value under field conditions.
- Limestone : A sedimentary rock composed primarily of calcite ($CaCO_3$) if dolomite ($CaCO_3.MgCO_3$) is present in appreciable quantities it is called a dolomitic limestone.
- Lithosequence : A group of related soil that differ one from the other in certain properties primarily as a result of parent material as a soil forming factor.
- Liquid Fertilizers : Fluid fertilizers that contain essential elements in liquid forms either as soluble nutrients as liquid suspensions or both.
- Loam : The textural class name for soil having a moderate amount of sand silt and clay loam. Soil contain 7 to 27% clay 28 to 50% silt and 23 to 52% sand.
- Loamy : Intermediate in texture and properties between fine textured and coarse textured soils includes all

textural classes with the words loam or loamy as a part of the class name such as clay loam or loamy sand.

- Luxury Consumption : The intake by a plant of an essential nutrient in amounts exceeding what it needs for example if potassium is abundant in the soil alfalfa may take in more than it requires.
- Lysimeter : A device for measuring percolation and evapotranspiration losses from a column of soil under controlled conditions.
- Macronutrient : A chemical element necessary in large amount (usually 50 mg/kg in the plant) for the growth of plants includes C, H, O, N, P, K, Ca, Mg and S (macro refers to quantity and not to the essentiality of the element).
- Macropores : Larger soil pores generally having a diameter greater than 0.06 mm from which water drains readily by gravity.
- Mass Flow : Movement of nutrients with the flow of water to plant roots.
- Metric Soil : A soil with developed soil horizons produced by the natural processes of soil formation and essentially in equilibrium with its present environment.
- Medium Texture : Intermediate between fine textured and coarse textured (soils) it includes the following textural classes very fine sandy loam, loam silt, loam and silt.
- Metric Potential : That portion of the total soil water potential due to the attractive forces between water and soil solids as represented through adsorption and capillarity it will always be negative.
- Maximum Retentive Capacity : The average moisture content of a disturbed sample of soil 1 cm. high which is at equilibrium with a water table at its lower surface.

- Mesophilic : Pertaining to moderate temperatures in the range of 15 to 35°C the range in which mesophilic organisms grow best and in which mesophilic composting takes place.
- Metamorphic Rock : A rock that has been greatly altered from its previous condition through the combined action of heat and pressure for example marble is a metamorphic rock produced from limestone gneisis is produced from granite and state is produced from shale.
- Methane (CH_4) : An odorless colourless gas commonly produced under anaerobic condition when released to the upper atmosphere methane contributes to global warming.
- Micas : Primary aluminosilicate minerals in which two silica tetrahedral sheets alternate with one alumina/ magnesia octahedral sheet with entrapled potassium atoms fitting between sheets they separate readily into visible sheets or flakes.
- Micronutrient : A chemical element necessary in only extremely small amounts (< 50 mg/kg in the plants) for the growth of plants examples are B, Cl, Cu, Fe, Mn and Zn (micro refers to the amount used rather than to its) essentially.
- Mineralization : The conversion of an element from an organic form to an inorganic state as a result of microbial decomposition.
- Mineral Soil : A soil consisting predominantly of and having its properties determined predominantly by mineral matter usually contains < 20% organic matter but may contain an organic surface layer up to 30 cm. thick.
- Moderately Coarse Texture : Consisting predominantly of coarse particles in soil textural classification it includes all the sandy loam except the very fine sandy loam.

- Moderately Fine Texture : Consisting predominantly of intermediate sized (soil) particles or with relatively small amounts of fine or coarse particles in soil textural classification it includes clay loam. Sandy loam sandy clay loam.
- Moisture Equivalent : The weight percentage of water retained by a previously saturated sample of soil 1 cm. in thickness after it has been subjected to a centrifugal force of 1000 time gravity for 30 min.
- Mollisols : An order in soil Taxonomy soils with nearly black organic-rich surface horizons and high supply of bases they have mollic epipedons and base saturation greater than 50% in any cambic or argillic horizon they lack the characteristics of vertisols and must not have oxic or spodic horizons.
- Montmorillonite : An aluminosilicate clay mineral in the smectite group with a 2:1 expanding crystal lattice with two silicon tetrahedral sheets enclosing an aluminium sheet. isomorphous substitution of magnesium for some of the aluminium has occurred in the octahedral sheet considerable. Expansion may be caused by water moving between silica sheets of contiguous layers.
- Moraine : An accumulation of drift with an initial topographic expression of its own built within a glaciated region chiefly by the direct action of glacial ice. Examples are ground lateral recessional and terminal Moraines.
- Morphology Soil : The constitution of the soil including the texture structure consistence colour and other physical chemical and biological properties of the various soil horizons that make up the soil profile.
- Muck Soil : (a) A soil containing 20 to 50% organic

matter. (b) An organic soil in which the organic matter is well decomposed.

- Neutral Soil : A soil in which the surface layer at least to normal plow depth is neither acid nor alkaline in reaction in practice this means the soil is within the pH range of 6.6 to 7.3.
- Nitrate Depression Period : A period of time beginning shortly after the addition of fresh highly carbonaceous organic materials to a soil during which decomposer micro-organisms have removed most of the soluble nitrate from the soil solution.
- Nitrification : The biochemical oxidation of ammonium to nitrate predominantly by autotrophic bacteria.
- Nitrogen Assimilation : The incorporation of nitrogen into organic cell substances by living organisms.
- Nitrogen Cycle : The sequence of chemical and biological changes undergone by nitrogen as it moves from the atmosphere into water soil and living organisms and upon death of these organisms (plants and animals) is recycled through a part or all of the entire process.
- Nitrogen Fixation : The biological conversion of elemental nitrogen (N_2) to organic combinations or to forms readily utilized in biological processes.
- Nonhumic Substances : The portion of soil organic matter comprised of relatively low molecular weight organic substances mostly identifiable biomolecules.
- O-Horizon : Organic horizon of mineral soils.
- Octahedral Sheet : Sheet of horizontally linked octahedral shaped units that serve as the basic structural components of silicate (clay) minerals each unit consists of a central six coordinated metallic atom (e.g., Al, Mg, or Fe) surrounded by six hydroxyl groups that in turn

are linked with other nearby metal atoms thereby serving as inter unit linkages that hold the sheet together.

- Order Soil : The category at the highest level of generalization in soil Taxonomy the properties selected to distinguish the orders are reflections of the degree of horizon development and the kinds of horizons present.
- Organic Fertilizer : By product from the processing of animal or vegetable substances that contain sufficient plant nutrients to be of value as fertilizers.
- Organic Soil : A soil in which more than half of the profile thickness is comprised of organic soil materials.
- Organic Soil Material (As used in soil Taxonomy) : (a) Saturated with water for prolonged periods. Unless artificially drained and having 18% or more organic carbon (by weight) if the mineral fraction is more than 60% clay more then 12% organic carbon if the mineral fraction has no clay or between 12 and 18% carbon if the clay. Content of the mineral fraction is between 0 and 60%. (b) Never saturated with water for more than a few days and having more than 20% organic carbon histosols develop on these organic soil materials there are three kinds of organic materials.
- Fabric Materials : The least decomposed of all the organic soil materials. Containing very high amounts of fibre that are well preserved and readily identitiable as to botanical origin.
- Hemic Materials : Intermediate in degree of decomposition of organic materials between the less decomposed fabric and the more decomposed sapric materials.
- Sapric Materials : The most highly decomposed of the organic materials having the highest bulk density least amount of plant fibre and lowest water content at saturation.

- Ortstein : An indurated layer in the B-horizon of spodosols in which the cementing material consists of illuviated sesquioxides (mostly iron) and organic matter.
- Oxidation : The loss of electrons by a substance, therefore a gain in positive valence charge and in some cases the chemical combination with oxygen gas.
- Oxisols : An order in soil Taxonomy soils with residual accumulations of low activity clays free oxides kaolin and quartz they are mostly in tropical climates.
- Parent Material : The unconsolidated and more or less chemically weathered mineral or organic matter from which the solum of soils is developed by pedogenic processes.
- Particle Density : The mass per unit volume of the soil particles in technical work usually expressed as metric long per cubic metre (mg/m^3) or grams per cubic centimetre (g/m^3).
- Particle Size : The effective diameter of a particle measured by sedimentation sieving or micrometric methods.
- Particle Size Analysis : Determination of the various amounts of the different separates in a soil sample usually by sedimentation sieving micrometry or combinations of these methods.
- Particle Size Distribution : The amounts of the various soil separates in a soil sample usually expressed as weight percentages.
- Particulate Organic Matter : A microbially active fraction of soil organic matter consisting largely of fine particles of partially decomposed plant tissue.
- Pascal : An SI unit of pressure equal to newton per square metre.

- Peat : Unconsolidated soil material consisting largely of undecomposed or only slightly decomposed. Organic matter accumulated under conditions of excessive moisture.
- Peat Soil : An organic soil containing more then 50% organic matter used in the united states to refer to the stage of decomposition of the organic matter peat referring to the slightly decomposed or undecomposed deposits and muck to the highly decomposed materials.
- Ped : A unit of soil structure such as an aggregate crumb prism block or granule formed by natural processes (in contrast to a clod which is formed artificially).
- Pedology : The science that deals with the formation of Morphology and classification of soil bodies as land scape, components.
- Pedoturbation : Physical disturbance and mixing of soil horizons by such forces as burrowing animals faunal pedoturbation or frost churning (cryoturbation).
- Penetrometer : An instrument consisting of rod with a cone shaped tip a means of measuring the force required to push the rod into a specified increment of soil.
- Percolation Soil Water : The downward movement of water through soil especially the downward flow of water in saturated or nearly saturated soil at hydraulic gradients of the order of 1.0 or less.
- Petrogypsic Horizon : A diagnostic sub-surface horizon that is a continuous strongly cemented massive gypsic horizon that is cemented by calcium sulphate it can be chipped with a spade when dry fragments do not slake in water and it is impenetrable to roots.
- pH Soil: The negative logarithm of the hydrogen ion activity (concentration) of a soil. The degree of acidity

(or a lakalinity) of a soil as determined by means of a glass or other suitable electrode or indicator at a specified moisture. Content of soil to water ratio and expressed in terms of the pH scale.

- pH Dependent Charge : That portion of the total change of the soil particles that is affected by and varies with changes in pH.
- Physical Properties (of soil) : Those characteristics processes or reaction of a soil that are caused by physical forces and that can be described by or expressed in physical terms or equation examples of physical properties are bulk density water holding capacity hydraulic conductivity porosity pore size distribution and so on.
- Plastic Soil : A soil capable of being molded or deformed continuously and permanently by relatively moderate pressure into various shapes.
- Platy : Consisting of soil aggregates that are developed predominantly along the horizontal axis laminated flaky.
- Plinthite (Brick) : A highly weathered mixture of sesquioxides of iron and aluminium with quartz and other diluents that occurs as red mottles and that changes irreversibly to hard pan upon alternate watting and drying.
- Plowing: A primary broad base tillage operation that is performed to shatter soil uniformly with partial to complete inversion.
- Point of Zero Charge : The pH value of a solution that in equilibrium with a particle whose net charge from all sources is zero.
- Polypedon (As used in soil Taxonomy) : Two or more contiguous pedons all of which are within the defined

limits of a single soil series commonly referred to as a soil individual.

- Porosity Soil : The volume percentage of the total soil bulk not occupied by solid particles.
- Potential Acidity : The acidity that could potentially be formed if reduced sulphur compounds in a potential acid sulphate soil water to become oxidized.
- Primary Mineral : A mineral that has not been altered chemically since deposition and crystallization from molten lava.
- Primary Effect : The increased decomposition of relatively stable soil humus under the influence of much enhanced generally biological activity resulting from the addition of fresh organic materials to a soil.
- Productivity Soil : The capacity of a soil for producing a specified plant or sequence of plants under a specified of plants or under a specified system of management. Productivity emphasizes. The capacity of soil to produce crops and should be expressed in terms of yields.
- Profile Soil : A vertical section of the soil through all its horizons and extending into parent material.
- Protonation : Attachment of protons (H^+ ions) to exposed OH groups on the surface of soil particles resulting in an overall positive charge on the particle surface.
- Puddled Soil : Dense massive soil artificially compacted when wet and having no aggregated structures the condition commonly results from the tillage of a clayey soil when it is wet.
- Reaction Soil (No longer used in soil science) : The degree of acidity or alkalinity of a soil usually expressed as a pH value or by terms ranging from extremely acid for pH values < 4.5 to very strongly alkaline for pH $>$ 9.0.

- Recharge Area : A geographic area in which an otherwise confined aquifer is exposed to surfical percolation of water to recharge the ground water in the aquifer.
- Redoximorphic Features : Soil properties associates with wetness that result from reduction and oxidation of iron and manganese compounds after saturation and desecration with water.
- Reduction : To gain of electrons and therefore the loss of positive valence. Charge by a substance in some case a loss of oxygen or a gain of hydrogen is also involved.
- Regolith : The unconsolidated mantle of weathered rock and soil material on the earth surface loose earth materials above soil rock (Approximately) equivalent to the term soil as use by many engineers.
- Relief : The relative differences in elevation between the upland summits and the lowlands or valleys of a given region.
- Residual Material : Unconsolidated and partly weathered mineral materials accumulated by disintegration of consolid rock in place.
- Rill : A small intermittent water course with steep sides usually only a few centimeters deep and hence, no obstacle to tillage operations.
- Rill Erosion : An erosion process in which numerous small channels of only several centimeters in depth are formed occurs mainly on recently cultivated soil.
- Rock : The material that forms the essential part of the earths solid crust including loose incoherent masses. Such as sand and gravel as well as solid masses of granite and limestone.
- Runoff : The portion of the precipitation on an area that is discharged from the area through stream channels

that which is lost without entering the soil is called surface runoff and that which enters the soil before reaching the stream is called groundwater runoff or seepage flow from groundwater (in soil science runoff usually refers to the water lost by surface flow in geology and hydraulics runoff usually includes with surface and subsurface flow).

- Salic Horizon : A diagnostic sub-surface horizon of enrichment with secondary salts more. Soluble in cold water than gypsum a salic horizon is 15 cm. or more in thickness.
- Sand : A soil particle between 0.05 and 2.0 mm in diameter a soil textural class.
- Saline Soil : A non-sodic soil containing sufficient soluble salts to impair its productivity the conductivity of a saturated extract is > 4 ds/m the exchangeable sodium adsorption ratio is less than about 13 and the pH is < 8.5.
- Saline Sodic Soil : A soil containing sufficient exchangeable sodium to interfere with the growth of most crop plants and containing appreciable quantities of soluble salts the exchangeable sodium adsorption ratio is > 13 the conductivity of the saturation extract is > 4 ds/m (at 25°C) and the pH is usually 8.5 or less in the saturated soil.
- Salinization : The process of accumulation of salts in soil.
- Saltation : Particle movement in water or wind where particles skip or bounce along the stream bed or soil surface.
- Saturated Paste Extract : The extract from a saturated soil paste the electrical conductivity E_c of which gives an indirect measure of salt content in a soil.
- Secondary Mineral : A mineral resulting from the

decomposition of a primary mineral or from the reprecipitation of the products of decomposition of a primary mineral.

- Sediment : Transported and deposited particles or aggregates derived from soil rock or biological materials.
- Sedimentary Rock : A rock formed from materials deposited from suspension or precipited from solution and usually being more or less. Consolidated the principal sedimentary rocks are sand stones shales limestones and conglomerates.
- Seed Bed : The soil prepared to promote the germination of seed and the growth of seedlings.
- Self Mulching Soil : A soil in which the surface layer becomes so well aggregated that it does not crust and seal under the impact of rain but instead serves as a surface mulch upon drying.
- Separate Soil : One of the individual sized groups of mineral soil particles sand silt or clay.
- Series Soil : The soil series is a sub-division of a family in soil Taxonomy and consists of soils that are similar in all major profile characteristics.
- Sheet Erosion : The removal of a fairly uniform layer of soil from the land surface by runoff water.
- Sheet (Mineralogy) : A flat array of more than one atomic thickness and composed of one or more levels of linked co-ordination polyhedra a sheet is thicker than a plan and thinner than a layer examples tetrahedral sheet octahedral sheet.
- Short Range Order Minerals : Minerals such as allophane whose structural framework consists of short distances of well ordered crystallin structure interspersed with distances of non-crystalline amorphous materials.

- Silica/Alumina Ratio : The molecules of silicon dioxide (SiO_2) per molecule of aluminium oxide (Al_2O_3) in clay minerals or in soils.
- Silt : (a) A soil separate consisting of particles between 0.05 and 0.002 mm in equivalent diameter. (b) A soil textural class.
- Silica/Sesquioxide Ratio : The molecules of silicon dioxide (SiO_2) per molecule of aluminium oxide (Al_2O_3) plus ferric oxide (Fe_2O_3) in clay minerals or in soils.
- Silting : The deposition of water born sediments in stream channels lakes reservoles or on floodplains usually resulting from a decrease in the velocity of the water.
- Slope : The degree of deviation of a surface from horizontal measured in a numerical ratio percent or degrees.
- Sodic Soil : A soil that contains sufficient sodium to interfere with the growth of most crops plants and in which the sodium adsorption ratio is 13 or greater.
- Sodium Adsorption Ratio (SAR):

$$SAR = \frac{[Na^+]}{\sqrt{\frac{1}{2}[Ca^{2+}]+[Mg^{2+}]}}$$

Where the cation concentrations are in milli moles of charge per litre (M mole/L).

- Soil : (a) A dynamic natural body composed of mineral and organic solids, gases, liquids, and living organisms which can serve as a medium for plant growth. (b) The collection of natural bodies occupying parts of the earths surface that is capable of supporting plant growth and that has properties resulting from the integrated effects of climate and living organisms acting upon parent material as conditioned by topography over periods of time.

- Soil Arc : The soil atmosphere the gaseous phase of the soil being that volume not occupied by soil or liquid.
- Soil Alkalinity : The degree a intensity of alkalinity of a soil expressed by a value > 7.0 on the pH seals.
- Soil Amendment : Any material such as lime gypsum sawdust or synthetic conditioner that is worked into the soil to make it more amendable to plant growth.
- Soil Association : A group of defined and named taxonomic soil units occurring to gather in an individual and characteristic pattern over a geographic region comparable to plant association in many ways.
- Soil classification (Soil Taxonomy) : The systematic arrangement of soils into group or categories on the basis of their characteristics.
- Soil Complex : A mapping unit used in detailed soil surveys where two or more defined taxonomic units are so intimately intermixed geographically that it is undesirable or impractical because of the scale being used to separate them a more intimate mixing of smaller areas of individual taxonomic units than that described under soil association.
- Soil Conservation : A combination of all management and land use methods that safeguard the soil against depletion or deterioration caused by nature and/or humans.
- Soil Correlation : The process of defining mapping naming and classifying the kinds of soil in a specific soil survey area the purpose being to ensure that soils are adequately defined accurately, mapped and uniformly named.
- Soil Geography : A sub-specialization of physical geography concerned with the areal distributions of soil types.

- Soil Loss Tolerance (T) : The maximum rate of annual soil loss that will permit plant. Productivity to be maintained the economically and indefinitely.
- Soil Management : The sum total of all tillage operations cropping practices fertilizer lime and other treatments conducted on or applied to a soil for the production of plants.
- Soil Map : A map showing the distribution of soil types or other soil mapping units in relation to the prominent physical and cultural features of the earth's surface.
- Soil Mechanics and Engineering : A sub-specialization of soil science concerned with the effect of forces on the soil and the application of engineering principles to problems involving the soil.
- Soil Monolith : A vertical section of soil profile removed from the soil and mounted for display or study.
- Soil Morphology : The physical constitution particularly the structural properties of a soil profile as exhibited by the kinds thicknesses and arrangement of the horizons in the profile and by the texture structure consistence and porosity of each horizon.
- Soil Organic Matter : The organic fraction of soil that includes plant and animal residues at various stages of decomposition cells and tissues of soil, organisms and substances synthesized by the soil population. Commonly determined is the amount of organic material contained in a soil sample passed through a 2 mm. sieve.
- Soil Quality : The capacity of a specific kind of soil to function within natural or managed ecosystem boundaries to sustain plant and animal productivity maintain or enhance water and air quality and support

human health and habitation sometimes. Considered in relation to this capacity in the undisturbed natural state.

- Soil Salinity : The amount of soluble salts in a soil expressed in terms of percentage milligrams per kilogram parts per million ppm or other convenient ratios.
- Soil Solution : The aqueous liquid phase of the soil and its solutes consisting of ions dissociated from the surface of the soil particles and of other soluble materials.
- Soil Strength : A transmit soil property related to the soil's solid phase conversion and adhesion.
- Soil Structure : The combination or arrangement of primary soil particles into secondary particles units or peds these secondary units may be but usually are not arranged in the profile in such a manner as to give a distinctive characteristic pattern; the secondary units are characterized and classified on the basis of size shape and degree of distinctness into classes types and grades respectively.
- Soil Structure Classes : A grouping of soil structural units or peds on the basis of size from the very fine to very coarse.
- Soil Structure Grades : A grouping or classification of soil structure on the basis of inter-and intra-aggregate adhesion cohesion or stability within the profile four grades of structure designated from 0 to 3 are recognized structureless weak moderate and strong.
- Soil Structure Types : A classification of soil structure based on the shape of the aggregates or peds and their arrangement in the profile including platy prismatic columnar blocky sub-angular blocky granulated and crumb.

- Soil Survey : The systematic examination description classification and mapping of soil in an area soil surveys are classified according to the kind and intensity of field examination.
- Soil Temperature Classes : A criterion used to differentiate soil in soil Taxonomy mainly at the family level classes are based on mean annual soil temperature and an differences between summer and winter temperature at a depth of 50 cm.
- Soil Textural Class : A grouping of soil textural units based on the relative properties of the various soil separates. (Sand, silt and clay) these textural classes listed from coarsest to the finest in texture are sand loamy sandy sand, loam, loam silt, sandy, clay loam silty clay loam, sandy clay, silty clay and clay there are several subclasses of the sand loam sand and sandy loam classes based on the dominant particle size of the sand fraction (e.g., loamy fine sand coarse sandy loam).
- Soil Texture : The relative proportions of the various soil separates in a soil.
- Solarization : The process of heating a soil in the field by covering it with clear plastic sheeting during sunny conditions the heat is meant to partially sterilize the upper 5 to 15 cm. of soil to reduce pest and pathogen populations.
- Solum : The upper and most weathered part of the soil profile the A and B horizon.
- Sombric Horizon : A diagnostic sub-surface horizon that contains illuvial humus but has a low cation exchange capacity and low percentage base saturation mostly restricted to cool moist soil of high plateaus and mountainous area of tropical and sub-tropical regions.

- Sorption : The removal from the soil solution of an ion or molecule by adsorption and absorption this term is often used when the exact mechanism of removal is not known.
- Species Diversity : The variety of different biological species present in an ecosystem generally high diversity is marked by many species with few individuals in each.
- Specific Heat Capacity : The amount of kinetic (heat) energy required to raise the temperature of 1 g of a substance. (Usually in reference to soil or soil components).
- Specific Surface : The solid particle surface area per unit mass or volume of the solid particles.
- Splash Erosion : The spattering of small soil particles caused by the impact of raindrops on very wet soils the loosend and separated particles may or may not be subsequently removed by surface runoff.
- Spodic Horizon : A diagnostic sub-surface horizon characterized by the illuvial accumulation of amorphous materials composed of aluminium and organic carbon with a without iron.
- Spodosols : An order in soil Taxonomy soil with ion sub-surface illuvial accumulations of organic matter and compounds of aluminium and usually land. These soils are formed in acid mainly coarse textured materials in humid and mostly cool or temperate climates.
- Stubble Mulch : The stubble of crops of crop residues left essentially in place on the land as a surface cover before and during the preparation of the seedbed and at least partly during the growing of a succeeding crop.
- Sub-group Soil : The great groups in soil Taxonomy are sub-divided into central concept sub-groups that show

the central properties of the great group intergrade sub-groups that show properties of more than one great group and other sub-groups for soils with a typical properties that are not characteristic of any great group.

- Sub-order Soil : A category in soil Taxonomy that narrows the ranges in soil moisture and temperature regions kinds of horizons and composition according to which of these is most important.
- Sub-soiling : Breaking of compact subsoils without inverting them with a special knife like instrument (chisel) which is pulled through the soil at depths usually of 30 to 60 cm. and at spacing usually of 1 to 2 cm.
- Sulphuric Horizon : A diagnostic sub-surface horizon in either mineral or organic soil that has a pH <3.5 and fresh straw coloured mottles (called jarosite mottles) forms by oxidation of sulphide rich materials and is highly toxic to plants.
- Surface Seal : A thin layer of fine particles deposited on the surface of a soil that greatly reduces the permeability of the soil surface to water.
- Surface Soil : The uppermost part of the soil ordinarily moved in tillage or its equivalent in uncultivated soils ranges in depth from 7 to 25 cm frequently designated as the plow layer the Ap layer or the Ap horizon.
- Surface Tension : The elastic like phenomenon resulting from the unbalanced attractions among liquid molecules (usually water) and between liquid and gaseous molecules (usually air) at the liquid gas interface.
- Taxonomy Soil : The science of classification of soils laws and principles governing the classifying of soil also a specific soil classification system developed by the US department of agriculture.

- Tensiometer : A device for measuring the negative pressure (or tension) of water in soil in situ a porous. Permeable ceramic cup connected through a tube to a manometer or vacuum gauge.
- Terrace : (a) A level usually narrow plain bordering a river lake or the sea rivers sometimes are boardered by terraces at different levels. (b) A raised more or less level or horizontal strip of earth usually constructed on a nearly on a contour and to designed to make the land suitable for tillage and to prevent accelerated erosion by diverting water from undesirable channels of concentration some times called diversion.
- Tetrahedral Sheet : Sheet of horizontally linked tetrahedron shaped units that serve as one of the basic structural components of silicate clay minerals each unit consist of a central four oxygen atoms that in turn are linked with other nearly atoms (e.g., Si, Al, Fe) thereby serving as interunit linkages to hold the sheet together.
- Thermal Analysis (Differential thermal analysis) : A method of analyzing a soil sample for constituents based on a differential rate of heating of the unknown and standard samples when a uniform source of heat is applied.
- Thermophilic : Peraining to temperatures in the range of 45 to 90°C the range in which thermophilic organisms grow best and in which thermophilic composting takes place.
- Thermic : A soil temperature class with mean annual temperature 15 to 22°C.
- Thixotrophy : The property of certain clay soil of becoming fluid when formed or agitated and then setting again when at rest similar to quick as in quick clays or quicksand.

- Till Drain : Pipe mode of burned clay concrete or ceramic material in short lengths usually laid with open joints to collect and carry excess water from the soil.
- Tillage : The mechanical manipulation of soil for any purpose but in agriculture it is usually restricted to the modifying of soil conditions for crop production.
- Tillage Primary : Tillage that contributes to the major soil manipulation commonly with a plow.
- Tillage Conservation : Any tillage sequence that reduces loss of soil or water relative to conventional tillage which generally leaves at least 30% of the soil surface covered by residues including the following systems.
- Minimum Tillage : The minimum soil manipulation necessary for crop production or meeting tillage requirements under the existing soil and climatic conditions.
- Mulch Tillage : Tillage or preparation of the soil in such a way that plant residues or other materials are left to cover the surface also called mulch farming trash farming stubble mulch tillage and plowless farming.
- No Tillage System : A procedure whereby a crop is planted directly into a seedbed not tilled since harvest of the previous crop also called zero tillage.
- Ridge Till : Planting on ridges formed by cultivation during the previous growing period.
- Strip Till : Planting is done in a narrow strip that has been tilled and mixed leaving the remainder of the soil surface undisturbed.
- Tillage Conventional : The combined primary and secondary tillage operations normally performed in preparing a seedbed for a given crop grown in a given geographic area.
- Tillage Secondary : Any tillage operations following

primary tillage designed to prepare a satisfactory seedbed for planting.

- Tilth : The physical condition of soil as related to its ease of tillage fitness as a seedbed and its impedance to seedling emergence and root penetration.
- Toposequence : A sequence of related soil that differ one from the other primarily because of topography as a soil formation factor.
- Top Soil : (a) The layer of soil moved in cultivation. (b) Presumably fertile soil material used to top dress roadbanks gardens and lawns.
- Trioctahedral : An octahedral sheet of silicate clays in which the sites for the six co-ordinated metallic atoms are mostly filled with divalent cations such as Mg^{2+}.
- Trophic Level : Levels in a food chain that pass nutrients and energy from one group of organisms to another.
- Ultisols : An order in soil Taxonomy soil that are low in bases and have sub-surface horizons of illuvial clay accumulation they are usually moist but during the warm season of the year some are dry part of the time.
- Umbric Epipedon : A diagnostic surface horizon of mineral soil that has the same requirements as the mollic epipedon with respect to colour thinkness organic carbon content structure and P_2O_5 content but that has a base saturation of less than 50%.
- Universal Soil Loss Equation (USLE) : An equation for predicting the average annual soil loss per unit area per year A $\Rightarrow$ RKLSPC. Where R is the climatic erosivity factor (rainfall plus runoff) K is the soil erodibility factor L is the length of slope S is the percent slope P is the soil erosion practices factor and C is the cropping and management factor.

- Vermiculite : A : 2 : 1 type silicate clay usually formed from mica that has a high net negative charge stemming mostly from extensive isomorphous substitution of aluminium for silicon in the tetrahedral sheet.
- Vermicompost : Compost made by earthworms eating raw organic materials in moist aerated piles which are kept shallow to avoid heat buildup that could kill the worms.
- Vertisols : An order in soil Taxonomy clayey soil with high shrink swell potential that have wide deep cracks when dry most of these soils have distinct wet and dry periods throughout the year.
- Vesicles : (a) Unconnected voids with smooth walls. (b) Spherical structures formed inside root cortical cells by vesicular arbuscular mycorrhizal fungi.
- Virgin Soil : A soil that has not been significantly disturbed from its natural environment.
- Water Table : The upper surface of ground water or that level below which the soil is saturated with water.
- Water Table Perched : The surface of a local zone of saturation held above the main body of ground water by an impermeable layer of stratum usually clay and separated from the main body of ground water by an unsaturated zone.
- Water use Efficiency : Dry matter or harvested portion of crop produced per unit of water consumed.
- Watershed : All the land and water within the geographical confines of a drainage divide or surrounding ridges that separate the area from neighboring water sheds.
- Water Stable Aggregate : A soil aggregate stable to the action of water such as falling drops or agitation as in watsieving analysis.

- Weathering : All physical and chemical changes produced in rock at or near the earths surface by atmospheric agents.
- Wet Land : An area of land that has hydric soil and hydrophytic vegetation typically flooded for part of the year and forming a transition zone between aquatic and terrestrial systems.
- Wilting Point (Permanent wilting point) : The moisture content of soil on an over dry basis at which plants wilt and fail to recover their turgidity when placed in dark humid atmosphere.

8

Plant Structure

Plant Physiology

Absorption and Movement of Water in Plant

- Water is the main source of plant and is essential for the maintenance of life, growth and development.

Process of Imbibition

- This is the process of absorption of water by the particles of solid substances without forming a solution.
- Imbibition helps in the germination of seeds movement of water from cell to cell in the plant body and the ripening of seeds.

Process of Diffusion

- This is the movement of molecules of gases, liquids and solutes from the higher concentration to the lower concentration area.

- The rate and direction of diffusion are influenced by the temperature and pressure.

Process of Osmosis

- In the osmosis the movement of solvent molecules from a region of its high free energy to a region of its low free energy when the two are separated by a semipermeable membrane till a state of equilibrium is reached.
- Osmosis is of two types-endosmosis and exosmosis.
- Osmosis plays an important role in the absorption of water by the root-hairs, cell to cell movement of water, induction of turgidity, providing mechanical strength, induction of turgidity.
- Osmosis helps the plants in developing resistance to drought injury, movement of plant parts and opening and closing of stomata.

Process of Plasmolysis

- This is the contraction and separation of protoplasm from cell walls due to exosmosis.
- The plasmolysis process is utilized in daily life in the preservation of meat, jellies and food stuffs and in preventing the growth of plants in the cracks of walls.
- The total amount of water present in the soil is called holard.
- The water available to the plant is called chesard.
- The water which cannot be absorbed is called echard.
- The symptoms appearing in a plant or the plant part due to the scarcity of water are known as wilting. Wilting may be temporary or permanent.
- Plants absorb water by two distinct mechanisms which independently operate. These mechanisms are active absorption and passive absorption.

- Active absorption of water involves the utilization of metabolic energy. Water is absorbed by the activity of root hairs.
- Passive absorption of water accounts for about 98% of the total water uptake by a plant. This process does not involve the expenditure of metabolic energy.
- Root pressure, guttation and bleeding are example of active absorption in plants.
- Root pressure is the pressure developed in the xylem vessels as a result of metabolic activity of roots.
- Under conditions of water scarcity and low temperature the roots develop negative pressure.
- Guttation is the exudation of water drops from the edges or tips of the leaves.
- It occurs in the plants, growing under conditions favouring high root pressure and low transpiration, through the structure called hydathodes.

How Bleeding Happens ?

- It is the exudation of sap from the injured parts of the plants.
- The upward transport of water from the roots to the aerial parts of plants is called ascent of sap.

Coherent Nature

- Cohesion tension and transpiration pull theory as proposed by Dixon and Jolly.
- It is widely accepted theory to explain the mechanism of ascent of sap in plants.
- The loss of water in the form of vapours from the living tissues of aerial parts of the plants is called as transpiration.
- Approximately 98% of the water absorbed by the land plants is lost in transpiration.

- Three kinds of transpiration are distinguishable-stomatal, cuticular and lenticular.
- Stomatal transpiration accounts for about 90% of the total water loss from the plant by transpiration.
- The opening and closing of stomata is controlled by the size and shape of guard cells resulting from the changes in their turgor.
- Transpiration is influenced by environmental factors such as-light, temperature, atmospheric humidity, wind velocity etc. and internal factor also influenced of transpiration.
- Transpiration is a necessary evil because of its advantages and disadvantages.
- Antitranspirants are the chemical substances that reduce the transpiration rate without effecting the other metabolic activities of the plants.

Process of Photosynthesis

"Photosynthesis is the process by which the green chlorophyll containing cells of the plants manufacture the carbon compounds with the help of inorganic raw material such as CO_2 and water in the presence of sunlight."

- The atmosphere contains only about 0.03% CO_2 by volume.
- The earth surface receives about 650×10^{21} cal. of sunlight energy per year.
- It is estimated that only 1.2% of this total energy is utilized in photosynthesis.
- All green parts of the plants contain chloroplasts and leaves are the main sites of photosynthesis.
- Chloroplast pigments are of three types :

Chlorophylls

- This is green photosynthetic pigments.

Carotenoids

- This is yellow, orange or red coloured pigments and absorb primarily the violet to blue regions of the spectrum.

Phycobilin

- According to the modern views, photosynthesis is fundamentally an oxidation-reduction process in which H_2O is oxidised to O_2 and CO_2 is reduced to carbohydrates and the radiant energy of the sun is stored in the form of chemical energy.
- According to the modern concept, photosynthesis is completed in two distinct steps namely light reaction and dark reaction.
- In the light reaction, ATP and $NADPH_2$ are generated and the water molecules undergo splitting resulting in the evolution of molecular oxygen.
- Photosystem-I is associated with both cyclic and non-cyclic photophosphorylation and reduction of NADP to $NADPH_2$.
- While photosystem II is concerned with only non-cyclic photophosphorylation and photo-oxidation of water resulting in the evolution of molecular oxygen.
- Dark reaction is also known as black man reaction in honour of the scientist who reported it first.
- The biochemical reaction of the "dark reaction phase" also known as Calvin cycle takes place in the stroma part of the chloroplast where all the necessary enzymes are located.
- Dark reaction primarily consists of three steps :
  - (i) carboxylation
  - (ii) glycolytic reversal and reduction
  - (iii) regeneration of RuBP

- Photorespiration is the process of uptake of O_2 and production of CO_2 in light by photosynthesising tissues.
- Photorespiration is the common in C_3 plants.
- Photorespiration is also called as C_2 cycle because of 2-carbon intermediates.
- C_4 plants (Sugarcane, maize, sorghum etc.) Partially overcome the disadvantage of photorespiration.
- C_4 plants show a characteristic Kranz anatomy of the leaves.
- Parts of the plant such as the green tissues (leaves) which synthesis the carbohydrates are known as "Source".
- The parts of the plants such as non-green tissues where carbohydrates are utilized for maintenance growth and development are called as "Sink".

9

Role of Machines

Although agricultural engineering is a fairly new subject, its effect is being felt in India increasingly. In recent years it has contributed considerably to agricultural production in the country both directly and indirectly. The number of modern machines used in agriculture is a fairly good index of the progress made during the four Five-Year Plans. Roughly, more than 1,00,000 tractors, crawler-tractors and power-tillers are in use in India.

Similarly, the number of combines, self-propelled, trailed and mounted ones, has gone up to more than 1,500. The number of power-threshers may be in the vicinity of 5,00,000 and the number of both diesel and electric pumps must have crossed the million mark. And yet all this progress mostly made during the last 25 years is insignificant as compared with the use of modern agricultural machines in the developed countries, such as the USSR, the USA, Japan, the UK, Canada and Australia, and we have yet to cross a number of hurdles, but the progress made so far is quite encouraging and forms a basis for a leap forward. This development has, however, made two things clear :

(1) that the Indian farmers, though belonging to a less developed country, are quick in adopting new methods, techniques and machines, provided they are convinced of their utility, and

(2) that the Indian industries can manufacture as good modern machines as those manufactured elsewhere. The confidence created by these developments augurs well for taking bold steps in the future. And this in itself is no mean achievement.

Agricultural engineering, as defined by international institutions, comprises four main branches, namely:

(i) farm implements and power,

(ii) rural structures,

(iii) soil conservation, drainage and irrigation, and

(iv) rural electricity.

Because of the peculiar local conditions, the lack of development of electricity on a large scale, and the immediate utility of implements, agricultural engineering in India has come to mean more of agricultural implements and machinery. But, in fact, agricultural engineering is a very vast subject. And may be defined as the application of the knowledge, techniques and disciplines of various fields of engineering to the solution of problems arising in the fields of agriculture and rural living, with the object of reducing labour, improving agricultural productivity per worker, and raising the standard of living of the farmers and increasing the overall earnings per worker.

During the Second and the Third Five-Year Plans, agricultural engineering divisions were added to the departments of agriculture in the states and at the Centre. They undertook the work of hiring or selling tractors, oilengines, implements, pumps, etc. Several training institutions and workshops were started to train artisans and mechanics, and in every state a research cum testing and training centre for agricultural implements was established. The cumulative effect of all these steps has been to popularize mechanization and to prepare a sound base for it.

Then came an era of more intensive planning and re-organization on a wider scale. With the establishment of agricultural universities, several agricultural engineering colleges were established, offering graduate and post-graduate courses. In most of the states, where such colleges were established, they took over research, education and extension from the state departments of agriculture. The most important step in the reorganization has been the establishment of agro-industries corporations, one in each state, with the financial assistance of the Central Ministry of Agriculture. The main objectives of establishing these corporations have been to take over the supply and service functions of the departments of agriculture and to expand them. These agro-industries corporations have each a paid-up capital of 2 to 5 crores of rupees. They have taken over the existing government workshops also. The progress made by these agro-industries corporations is given in this chapter.

The other most important development has been in the field of industry which has established so far half a dozen factories for manufacturing tractors, 3 or 4 for manufacturing power-tillers and several other factories for manufacturing oil-engines and electrically-driven pumping-sets. Some of these industries are being substantially assisted by the Industrial Finance Corporation of India. The Government has also established a factory for manufacturing crawler tractors and another one for manufacturing wheel-type tractors. These three main developments helped to quicken the pace of mechanization of Indian agriculture.

The following tabulated statement clearly shows the progress made during the last 25 years:

Sl. No	Name of the Implements	1951	1956	1961
1.	Mouldboard-type iron ploughs	9,30,000	13,66,000	23,00,000
2.	Bullock-driven sugarcane crushers	5,04,000	5,44,000	5,90,000
3.	Power-driven sugarcane crushers	21,000	23,000	32,210
4.	Centrifugal pumps with oil-engines	82,000	1,22,000	2,29,000
5.	Electric pumps	25,000	55,000	1,60,000
6.	Tractors	8,500	21,000	34,000

It is not merely the knowledge of agricultural engineering, including implements and machinery, which is necessary for increasing agricultural production, but also the knowledge of the sources of their availability, prices and upkeep. The following is a list of sources through which the information on the availability and prices of the machinery can be obtained.

(1) At the Village Level: the Village-Level Worker or *gram sewak.*

(2) At the Block Level : the Agricultural Extension Officer or the District Planning Officer.

(3) At the District Level: the District Agricultural Officer or the District Planning Officer.

(4) At the State Level: the State Agricultural Engineering Section and the Managing Director of the State Agro-Industries Corporation.

(5) At the Central Government Level: The Director of Agricultural Implements and Joint Commissioner (Machinery), Ministry of Agriculture and Irrigation,

Krishi Bhavan, New Delhi. The private industry as well as the government, through their agro-industries corporations, are in the field now for the supply of agricultural machinery to the farmers. To prevent the charging of exorbitant prices for tractors, the government has fixed the maximum prices of tractors with the assistance of the Bureau of Cost Accounting in the Ministry of Finance. This organisation has helped the farmers very much.

Future Prospect : Agricultural aviation is now gaining ground in India. At present, the aeroplanes and helicopters are being used mostly for spraying pesticides, but in future, they may be used for emergency sowing, and spraying liquid fertilizers on the growing crops. Agricultural electronics has also begun and the remote-control tractors are being designed and tested in some countries. Japan has successfully designed machines for transplanting paddy seedlings and their trials are going on at the Tractor-Training and Testing Centre at Budni. A very ambitious programme for joining the Ganga with the Kauveri, via the Mahanadi, the Narmada, the Godavari and the Krishna, is under the consideration of the Central Government and some preliminary work is being carried on. With the successful explosion of an atomic device, India may be in a position to generate more electricity in future. The availability of plentiful irrigation water as a result of the Ganga-Kauveri Project, and the availability of more electricity from atomic energy will make the progress of farm mechanization in India speedier. Thus the future seems to be bright.

Agriculture and Mechanisation

The beginning of mechanization of Indian agriculture was made by the use of improved hand-tools and improved bullock-drawn implements. Particular stress was laid on these items during the first three Five-Year Plans and slowly more sophisticated implements were introduced during the Fourth and Fifth Plans. These included power driven threshers, tractors and the combine harvesters. Since simpler implements were taken up first and since in many areas they are still being used, their description is taken up here first. Quite a variety of implements and machinery is required for carrying out various agricultural operations, starting from the opening of the land to harvesting, threshing, winnowing and storage. These operations include ploughing, harrowing, levelling, sowing, intercultivation, the application of fertilizers and manures, harvesting, threshing and winnowing.

India is a sub-continent and a variety of indigenous or *desi* implements are in use in different parts of the country for the last so many centuries. Before efforts were made to introduce foreign implements, it was necessary that a survey of indigenous agricultural implements used in India should be undertaken. Therefore, under the Indian Council of Agricultural Research, a countrywide survey was carried out and its report was published. This survey has brought a number of important facts to light. The number of implements used by the Indian farmers is really very large. Some of these implements, even if they look crude, are really useful and the consensus of the agricultural engineers in India has been that it would be easier at least as a first step to improve upon the indigenous implements in order to increase their efficiency and thus give to the Indian farmers something which is within their means.

To achieve this objective effectively, the Indian Council of Agricultural Research has now established one research-cum-testing and training centre in each state. Some of them have now been converted into regional centres or have been taken over by the state governments. The object is to improve the indigenous agricultural implements, to design new implements and to test them in the field up to the prototype level. After the implements have proved to be useful, the prototypes can be given to the manufacturers both in the public as well as in the private sectors and the implements can then be sold through them to the farmers.

Research Principles : In India, the holdings are generally small, the average farmer is poor, and the draft cattle are also small and ill-fed. Taking these factors into consideration, certain principles, on which research is to be carried out, have been laid down. These are given below :

(1) That the implements and machines for the Indian farmers should be simple in construction so that they could be operated by the illiterate farmers, and should either be manufactured or could be repaired by the village artisans or mechanics.

(2) The prices of the implements should be within the reach of the ordinary farmers, though they need not necessarily be cheap at the cost of efficiency, but they should also not be too costly. In the case of power-driven equipment, the difficulty about the cost can be overcome by giving the machinery on hire or hire-purchase system or on custom basis or can be utilized on co-operative basis.

(3) The implements should be light, so that they can be transported from the village to the field and that the draft of the implement is suitable for the types of animals that are used in particular areas.

(4) That the implements should be scientifically tested on a fairly uniform basis throughout India, so that the results are known and can be compared. There should be as little guess work as possible.

(5) In designing the implements, as far as possible, the local available materials are to be used. In the case of soil-working and plant-working parts, durable materials, such as high-carbon steel, should be used.

To fulfil these requirements, the Indian Council of Agricultural Research has also finalized procedures for the testing and designing of implements and the proformas in which the test data are to be collected. In recent years, the attention of the Indian farmers has been drawn to the power-operated machines and equipments, such as tractors, oil-engines, and pumps. Quite a large number of the farmers are using these implements and firms have also been established in India to manufacture them. This trend towards mechanized farming is bound to increase.

Recently, however, new developments have taken place to bring about mechanization of agriculture. Owing to the shortage of diesel-oil and its increased price, the mechanical operation will have to be restricted to deep ploughing, land-levelling, clearing land, and other operations which cannot be carried out with bullock-power. Now greater attention has to be paid to the bullock drawn implements and manually operated implements. For instance, for short distances, bullock-carts have been found more economical than trucks; hence the necessity for improving bullock-carts and also bullock-drawn, but improved implements, on priority basis. The description of some improved implements is given in the following pages:

Ploughs, a Primary Implement

A plough is the primary implement used all over India. The *desi* ploughs have almost a common design and they differ mostly in size, depending upon the type of the soil and the size of the bullocks or buffaloes used in the particular area. The heaviest ploughs are found in the black-cotton-soil areas of India. The *desi* plough is almost like a western-style cultivator, both in design and action. The body and the wedge may be of one piece or of two pieces. The iron point or share is made up of a rod or bar of mild steel, and penetrates into the soil and protects the wooden wedge against wear and tear. The beam varies from 2.5 to 3.5 m in length and is made of wood.

The handle is fixed to the beam at the back of the body. All indigenous ploughs are almost similar in design, but vary considerably in size and weight to suit the prevailing soil types, ploughing conditions and the weights and heights of the draft animals in the locality. The working draft varies from 27 to 148 kg. The following table gives an idea of the size and the draft of the *desi* ploughs used in different states.

The *desi* plough makes a V-shaped furrow and, unlike a mouldboard plough does not turn over the soil. It is mostly made of hard wood, e.g. *babool* or *kikar* (*Acacia nilotica).* The depth of ploughing is usually from 10 to 2.0 cm and the width from 13 to 18 cm. There are minor variations in design, weight, the length of the beam, etc. The duck-foot type of plough, as used in Orissa is different in design. In has a flat bottom of about 38x30 cm and is said to be useful under soil conditions in Orissa. The plough used in Maharashtra is very heavy requiring sometimes 2 to 4 pairs of bullocks. It is useful for ploughing the black cotton soils of the Deccan. Most of the *desi* ploughs are light in weight and can be carried either on the yoke of the bullocks, or on the shoulders by the farmers.

The depth of the plough can be adjusted with a wedge provided in the body of the plough or by adjusting the yoke either forward or backward on the beam. If the yoke is fixed forward, the plough goes deeper and if the yoke is taken backward, the plough goes shallower. In Andhra Pradesh, a simple triangular plate is fixed in front of the body of the plough so that ridges and furrows can be made. This is an ingenious device to convert the *desi* plough into a ridger. Since the *desi* plough acts like a cultivator, repeated ploughings and cross ploughings are necessary to secure a good tilth. The *desi* ploughs are also used for making shallow furrows for depositing seeds, for inter cultivation between the rows of crops and to harvest crops such as potatoes, groundnuts and sweet-potatoes. It can thus be seen that the *desi* plough is a multipurpose implement difficult to be replaced, unless more efficient multipurpose improved implements are designed.

Weight, Draft and the Furrow Size of Different Desi Ploughs

Ploughs	*Weight (kg)*	*Length of bean (m)*	*Average draught (kg)*	*Horse power expended*	*Average size of furrow (cm)*	
					Width	*Depth*
Bareilly	10.88	2.66	28.57	0.40	17.78	10.16
Pusa (Bihar)	15.42	2.5	27.66	0.43	26.4	17.78
Angul (Orissa)	10.88	2.15	47.17	0.73	27.94	15.42
Sabour (Bengal)	19.95	2.87	100.24	1.35	30.4	20.32
Patna	15.87	2.71	54.43	0.81	27.94	20.32
Kerala	23.58	3.32	134.26	1.87	30.4	20.32
Dohad (Mumbai)	13.15	2.84	158.75	2.67	22.86	22.86
Coimbatore	37	3.53	328.0	1.71	27.94	22.00
Indore	57	3.15	403.5	1.71	22.00	26.00

Technique of Bladed Harrows

These are called *bakhars* or *guntakas. Bakhar* is a very common implement used in the Deccan. It is one of the best implements for primary tillage. The farmers use it for preparing the land and for sowing the seed. It consists of a steel blade of about 45 to 90 cm in length and 8 cm in width. It can cultivate the land to a depth of 9 to 10 cm. Even the dryland tillage is done with this implement by adding a wedge to it. If the soil is very dry, a man stands on the implement to give it the weight. It can cover 0.8 to 1.6 ha in a day because of its greater width. The Lanka *bakhar* used in the Vidarbha area has a blade even up to 1.5 m in length and is used after a few monsoon showers to kill young germinated weeds. The implement is worked cross-wise in the fields to achieve a better tilth. This is a very useful implement and costs only Rs 30 to 45. Somehow, this implement has not yet been introduced into areas above the Narmada river, except in some pockets near Jhansi and Gwalior.

It needs to be examined whether such an implement can be introduced into Bihar, Uttar Pradesh and Punjab. Almost every farmer in the Deccan possesses this implement. It is interesting to mention here that in some parts of the USA, where moisture conservation is an important problem, the use of "sweeps" and "chisel cultivators" which are similar to the Indian *bakhar* or blade harrow in working, is being advocated. The Egyptian and West Asian ploughs are also very much like the Indian bladed harrow. Recently, improvements have been made in the design of the *bakhars*, making it possible to change the blade as desired. They are mostly made of steel.

Relevance of Mouldboard in Improving Plough : A mouldboard plough consists of a share and a mouldboard of steel. This is probably one of the first implements to be introduced into India from foreign countries. It differs in action from the *desi* plough in ploughing a square furrow and in inverting the soil either completely or partially, depending on the curvature given to the mouldboard. For a proper functioning of the mouldboard ploughs, the provision of an adequate vertical and horizontal suction to the share up is necessary. In the case of long-beam ploughs, the beam and the handle are made of wood, whereas in the case of a short-beam plough, the complete implement is made of metal. Where it is necessary to have a one-way ploughing, the turnwrest type of ploughs are used. Lakhs of such ploughs are in use in Maharashtra.

There, the ploughs are now available on hire also. It is being adopted where green-manuring crops or surface vegetation have to be buried to hasten decomposition. Heavy mouldboard ploughs require sturdier animals, and sometimes two or three pairs of bullocks are used. They are also employed for the eradication of deep-rooted weeds, e.g. *kans (Saccharum spontaneum)* and *kunds (Cyperus rotundus)*. The turnwrest type of mouldboard ploughs are very efficient. For lighter ploughing, the beam-type ploughs are used. The shares of the lighter ploughs should be made of steel instead of cast iron. These ploughs are finding increased favour in several parts of northern

India and are replacing the *desi* plough. The mouldboard plough in common use is with bullock-power cut furrow slices varying from 12 to 23 cm in width. Typical figures in respect of their relative performances are given below :

Double Mould Boards : These ploughs have double mould-boards. There are three sizes—the light, the medium and the heavy. The heavy ones are used for the cultivation of sugarcane and the lighter ones for the cultivation of vegetables and other crops. Some of the improved ridger ploughs have an arrangement for adjusting the width of the furrow by changing the distance between the two mould boards. For vegetable orchards, and for farmers who grow sugarcane, potatoes, etc., the possession of the ridger is a must.

Width of the plough (cm)	-	17.78	16.31 (nominal)	10.43 (nominal)
Width of the furrow cut (cm)	-	17.78	13.34 to 16.31	9.16 to 10.43
Depth of the cut (cm)	-	12.7	12.7	12.7
Drought in dry sandy loam at Delhi (kg)	-	123.37	127	118
Drought per square inch of furrow cross-section (kg)	-	0.562	0.632	0.913
Speed of bullocks (km)		3	1.673	1.737

Harrows and Cultivators : The indigenous harrows and cultivators are known as *bakhars* and *guntakas.* They mostly consist of a long sharp blade fixed to a wooden implement. The body of such an implement is made of *babool* wood, cut into a rectangular or octagonal shape. The length of the body is 91 to 182 cm and the width is 25 to 30 cm. The blade is attached to the log on its underside with two slanting wooden pegs. These implements are used for preparing the seedbed, for covering the seed and also for summer cultivation, provided the soil is not too hard.

Improved *bakhars* and *guntakas* have also been, lately, manufactured. The improvements relate to the changing of the blades, the adjustment of the depth and the attachments that may be needed in the case of these harrows. The length of the blade is from 61 to 152 cm and, therefore, this implement can cover quite a large area in a day. Steel *bakhars* and *guntakas* are nowadays mass-manufactured in India. Their prices range from Rs 40 to 55.

Some of the harrows used in foreign countries, e.g. the spike-tooth harrow, peg-tooth harrow, the spring-tooth harrow, and disc harrows have also been introduced. Out of all these harrows, the disc harrows have proved to be very popular. The spike-tooth harrow is also often used, particularly in parts of Punjab and western Uttar Pradesh. The spike-tooth harrow is also often used for inter-cultivation and for thinning crops in Punjab. The triangular frame on which spikes are fixed is used in that state.

Use of Soil Implements

Soil for Scoop : A soil scoop consists of a shallow box-like structure with an open front. On its front, a sharp blade is fixed and two hooks are attached to it. The bullocks are yoked to the hooks. Two wooden handles are provided at the back. The capacity of the soil scoop is 0.056 to 0.084 cu m. This implement is used when the soil is to be carried to a fairly long distance. The price of the soil scoop is Rs 60 to 1,000.

Wheel Barrows: Common wheel barrows are very useful for gardens or orchards. The wheels may be of mild steel or of rubber. The price of a wheel barrow is Rs 150 to 200.

Keni Making or Levelling Karaha : It consists of a fiat blade 1 to 5 m long, with a vertical handle and two rings in the front for yoking the bullocks with ropes. This is probably the commonest implement used in soil conservation for making bunds and levelling the land. Literally, thousands of these implements are used in Ahmadnagar district of Maharashtra for bunding. The bunding of about 200,000 ha has been completed with this implement. It is also becoming popular in other parts of Maharashtra, Andhra Pradesh and Karnataka for bunding land as well as for bunding *nullahs*.

The space where the bund is to be formed is first marked, and on both sides of this space, a width of about 5 m is ploughed with a mouldboard plough. The loosened soil is then moved on to form a bund. If the land to be ploughed is hard and cloddy, it hurts the feet of the bullocks and reduces their life. As far as possible, the bunds should be formed when the soil is soft. The price of the *keni* is Rs 40 to 60.

Patela : The indigenous implement is called *patela* or *sohaga*. It is merely a flat log 1.8 to 2 m long and is dragged over the clods by bullocks. The driver usually stands on the log to put weight on it and hold it down.

In Uttar Pradesh, the improved *pata* or *patela* is now mass-manufactured and has become quite popular. It is called *Singh patela*. It is provided with a series of pointed hooks made of 10 mm to 15 mm rods of mild steel. These hooks help to crush the clods and to collect the weeds. This implement costs 100 to 150 rupees, depending upon the length of the wooden blade and the number of steel rods used.

A land-leveller or *jandra* is also used in some tracts. It consists of a plank of wood, about 1 to 1.5 m long, 12 cm and 2.5 cm thick. In those areas where bamboo is readily available, a ladder-shaped leveller made of bamboo is used.

Bunds Forming : For soil conservation, sometimes it is necessary to make temporary low-level bunds, 18 to 22 cm high. These bunds last for one agricultural season. They either get washed away or are obliterated by subsequent ploughing and cultivation. For forming such bunds, an implement, called bund-former, was originally developed at Coimbatore. The implement consists of two concave blades, 90 to 150 cm long and 10 to 15 cm wide, joined across in a conversing manner.

This implement is to be worked in a well-ploughed and harrowed field, just for collecting the soil and not for digging it. Its job is to collect the soil at the wider front and deposit it in the form of a ridge or a low bund behind. There are three sizes of bund-formers with blades 9 and 12 cm long. Recently, a bigger bund-former, called "sara-making" machine, has also been designed. A wooden bund-former that would cost less than 10 rupees has been prepared at the Agricultural Workshop at Nilokheri (Haryana). Its blades are made of *babool* wood and its construction is so simple that the implement can be easily made by a village carpenter. If the wood is made available by the farmer, the implement would not cost more than 6 rupees.

The Forest Department in Madras has developed basin-listers for forming a series of a small basin or depressions in the land in order to collect rain-water. This is a useful operation in dry farming, because it conserves moisture. However, the implement, being rather costly, has not become popular.

Seed-sowing Devices

Among the indigenous implements, many seed-sowing devices are attached to implements, such as the *desi* plough. They are commonly called *nari* plough, *tifan*, *argada* and other multi-type implements that are used in the Deccan. Tube or tubes are fixed to these implements and on the top of these, a seeding-bowl, in which the seed is dropped, is fixed. The seeding-bowl has a number of holes in it. The construction of the holes is ingenious, with the result that the seed is dropped fairly evenly into the tubes and goes down to the tubes. In some areas, which are considered backward agriculturally, broadcasting or the sowing of seed at random on the cultivated field is also followed but the major portion of the area is nowadays sown in line and the seed is drilled. Sowing in lines ensures the proper distribution and placing of seed, resulting in the uniform spacing of the plants, with a concomitant reduction in seed-rate. It also helps the seeds to germinate regularly, ensuring a better stand of the crop and also facilitates inter-culture with bullock-operated hoes, thus reducing the cost of cultivation.

Quite a large number of improved seed-drills, both automatic and hand-operated, have been designed. Among the hand-operated ones, a copy of the wooden *tifan* has been made in Uttar Pradesh. It is made of mild steel, with an arrangement for changing the distance between the tines. The price of such a seed-drill is Rs 75 to 100. Among the automatic seed-drills mention may be made of the McCormic seed-drill, the Master seed-drill, the Gunti seed-drill and the Kirloskar seed-drill. The McCormic seed-drill uses an external-force feed-type of seed-metering device. The price of this seed-drill for a three-tined implement is Rs 1,200. The Master seed-drill has a series of small cups or spoons fixed on the discs. These discs revolve in a box filled with seed. The spoons gather the seeds and drop them into the tubes. The price of the Master seed-drill is Rs 950. The Gunti seed-drill also uses an external-force feed-type seed-metering device, but it has also a fertilizer-metering box fixed alongside the seed-box. The Kirloskar seed-drill which has recently been introduced has an ingenious device for seeding but it has not become popular. Most of the manufacturers of seed-drills are now making efforts to combine the simultaneous feeding of fertilizer with the dropping of the seed, thus making the implement a seed cum fertilizer drill. So far, however, none of these combined devices has proved very popular because of high cost, complicated designs greater draft, etc.

The seed-drills need a lot of skill in operation. It is necessary, particularly in respect of automatic seed-drills to calibrate them before sowing, so as to know the exact seed-rate that it will drop. The seed-drills are to be calibrated for each type of crop separately because of the differences in the shape, size and weight of the seeds of different crops. Similarly, in the case of the automatic seed-drills, proper care has to be exercised in choosing the soil-working part, such as the tine. If the soil is hard, strong straight tines with sharp edges in front should be selected. If the soil is soft, the disc shovels may be preferred. Some other important factors which a designer of a seed-drill and its user have to keep in mind can also be mentioned. These are the covering of the seeds, the pressing of the soil slightly over the seeds, the transport facilities for these drills, etc. Among the sowing devices, mention can also be made of transplanting-machines. Even though strenuous efforts are being made in different countries, such as Japan, Italy, the USA and China, so far no really useful machine for transplanting paddy has been prepared, except in Japan. Some of these new machines have been brought for trial at the Central Tractor Testing and Training Station, Budni, and have given quite good results as far as transplanting is concerned, but the machines require specially raised seedlings in trays and this is an elaborate and costly process. Therefore, transplanting has to be done by hand. A small tool for transplanting two or three seedlings of paddy at a time is being used in Burma and Malaysia. There is need to introduce this tool in India, because with little practice it can increase the efficiency of the labourers at least by 20 per cent.

Most of the interculture in India is done with a hand-tool, called *khurpi,* which consists of a sharp-edged triangular or sickle-shaped blade. Sometimes, weeding is done with a small spade, hoe or *mummuty*. Most of these tools have small handles and the user has to sit on his heel? or bend his back while weeding.

Weeding is one single operation which can increase crop yields without much additional investment, because efficient weeding will save moisture, crop nutrients and, to some extent, air and light required for the cultivated crop. To increase the efficiency of weeding, long-handled hoes, such as the Singh Hand-hoe, the Sharma Hand-hoe and the Dutch hoe have been introduced. The price of the hand-hoe is around Rs 6, including the handle and it is well within the means of even a poor farmer. It has been observed that four to five times more of weeding can be done with such long-handled hoes than with a *khurpi.* These hand-hoes are simple in construction and can be made by the village blacksmith. The use of these hand-hoes is, therefore, strongly recommended.

The wheel-hoes are a further improvement upon the hand-hoes. A wheel-hoe consists of a body, handle and a wheel. To its body, different attachments, such as shovels, a small plough, a small ridger, and other types of blades can be fixed. The price of the wheel-hoe is Rs 60 to 75, with a Weeding-attachment. Other attachments can be purchased at extra cost. Being provided with a wheel, it is easier to work and is very useful for inter cultivation in vegetable gardens and orchards. Where the soil is hard, one man can pull this hoe and another can direct it. Where the soil is soft, it requires only one man to work it.

For hoeing line-sown crops with bullock power a light *desi* plough is used in northern India. In the area between the Narmada river and the Krishna river, multi-tined bullock-operated hoes are used. They are known as *doura (Gunti), douri, dundia* and *dulari.* These are the miniature forms of *bakhars.* When the plants are small, multi-tined hoes fixed to a single wooden body are also used in Andhra Pradesh and northern Karnataka. The draft of these hoes is very light and sometimes two to four of them are pulled by one pair of bullocks. In the dry farming areas of Maharashtra, a *doura* with a split blade is used. The blades in this case work on each side of the row of plants. In the case of rain-fed crops, hoeing by using bullocks is done at the most three or four times, after which the plants grow tall and bushy and also the soil becomes hard. To cope with this condition, stronger and sharp hoes have been designed. They are commonly known as the Washim hoe, the Akola hoe, the Baroda hoe and the *triphali.* The designs of these hoes are more or less similar. They consist of three tines with a sharp triangular blade fixed to each. These hoes are very useful for the interculture of the cotton crop sown in rows.

Yet another implement called Indore Ridger, was introduced

in Malwa some years back. This implement has a sharp beam, with a sloping ends and a hexagonal share attached to it with a strong iron bar. The ridger can be used up to a later stage of the crop without doing physical injury to the plants. It not only eradicates weeds but also produces a good mulch. Similar to the *doura,* two or three such ridgers can be attached to a pair of bullocks. It is surprising that such bullock-drawn inter-cultural implements have not yet been introduced into northern India despite their being used by the Indian farmers in the Deccan for centuries. There is a great deal of scope to introduce such implements into those areas where they are not being used at present.

For some special crops, such as paddy, special types of paddy-weeders have been designed and are being mass-manufactured in India. They are based on the Japanese design. They consist of two rollers with pointed fingers and are pushed in between the rows of the paddy plants when water is standing in the field. The idea of using these weeders is not to uproot the weeds and bring them to surface, as in the case of other weeders, but to bury them in the mud so as to suffocate and decompose them, and add organic matter to the soil.

With the advancement of scientific research, new methods to control weeds with selective weedicides, such as 2, 4-D, are being used. When this weedicide is sprayed in between the rows of a crop it kills the weeds only and not the crop. Sprayers are used for spraying these weedicides. At present, the weedicides are costly, but when their supply becomes abundant, the killing of the weeds with selective weedicides may become quite popular in India and it may not be necessary to use other implements to a great extent.

The other implements may have to be used for forming a fine mulch in between the rows, so as to conserve soil moisture. Some of the intercultural implements may have some attachments fixed to them to carry out some special jobs, such as sulphur-dusting in tobacco cultivation and the top-dressing of fertilizers. Already, simple sulphur dusters which can be worked in rows, and fertilizer attachments to the weeders have been designed.

Harvesting Implements

So far, the harvesting operation has not yielded to any mechanized tool. Almost universally, sickle is the only tool that is used for harvesting crops in India even though efforts have been made to introduce reapers, mowers, and even combined harvesters and threshers. These machines have, however, proved successful only on big estates or government farms, but because of their high cost, they have not become popular with the Indian farmers except when used on custom basis, as in Punjab and in Haryana during the last five years. The European scythe has also been tried, but except for cutting of lucerne or some such green fodder they have not proved very useful. One of the main reasons seems to be the shattering of the grain because of the impact of the harvesting implements and machines. In the case of sickles, there is less of shattering. The sickles are of two types, namely, plain or serrated or saw-shaped. The serrated sickles are used for cutting paddy and wheat.

Thrashing Implements

Threshing in, the case of paddy, if the area is small, is done mostly by beating the harvested plants with sticks. In the case of other crops, it is done by treading the crop with a team of animals. Stone rollers, about 90 cm in length and 30 to 60 cm in diameter, have also been used in some areas. However, no other thresher has become so popular as the "Olpad Thresher", which consists of 14 to 21 serrated discs fixed on a rectangular wooden frame in three lines. A seat is provided for the driver and also a guard. The implement is pulled by a pair of bullocks and worked on the harvested material which is spread-thickly on the threshing-floor. Threshing with this thresher is not only cheaper but is also quicker and more efficient than treading by bullocks and by using other methods.

The Japanese pedal-operated thresher, which was introduced into the rice-growing areas has not proved very popular because it does not give as much yield of grain as beating with sticks does. The reason for the Japanese using this type of thresher is that they are keen to get as long pieces of straw as possible in order to prepare ropes, mats or sacks out of the paddy straw. These articles are not used by the Indian farmers and, therefore, this thresher has not become popular in India. It can thresh about 300 kg of paddy in a working day of eight hours.

The threshing of wheat in some of the large private and government farms is done by trampling the harvested crop under a tractor or under a disc harrow attached to a tractor. It has been noticed that a 15-bhp tractor with a suitable harrow produces about 1,860 to 2,975 kg of clean undamaged grain per day of 10 hours at a working cost of Rs 35 to 40. The *bhusa* or the broken straw is also of good quality.

Recently, the Government of Punjab has introduced a power-operated wheat-thresher. It is worked with a tractor or an oil-engine of 20 to 25 bhp. This thresher is being simplified so that it can also be operated with an oil-engine of 10 to 15 bhp. The thresher is used by the owner alone if he himself is a big farmer, or can be given to other farmer, or can be given to other farmers on custom basis, or can be used co-operatively. In most of the villages in Punjab, Haryana, and western Uttar Pradesh, now at least one thresher exists, and in many cases there is more than one. Thousands are used on custom basis. The *bhusa* is a precious commodity and finds a ready market as a cattle feed. Therefore, a thresher, which cannot make bhusa, will not become popular. At the research stations in the country, the Japanese power-operated paddy-thresher is being given a trial. This machine can turn out 50 kg of clean paddy per hour. However, it has no *bhusa-making* arrangement. To make *bhusa*, the straw has again to be trampled under the feet of the bullocks. The drum-type power-operated thresher of 10 to 15 bhp has become very popular in northern India. This thresher is an example of a machine which the farmers in that region have readily adopted because of its great utility. Now virtually threshing by trampling straw by using, bullocks is limited to very small farmers. The introduction of this machine has saved much labour and has made threshing very easy and to complete it before the monsoon.

Winnowing-fans and Winnower Tools in Different Types: To separate the chaff from the grain, the threshed material is put before a good natural breeze and the grain is allowed to drop through gravity and be separated from *the bhusa*, which being lighter, falls a little further away from the grain heap. In many areas, however, the natural breeze at the time of winnowing is not readily available and the farmers have to wait for hours and hours even for days to avail themselves of the natural breeze. Usually, in such areas, there is a danger of rain and hail-storm during this period. The farmers are, therefore, very anxious to take the season's produce to their homes as early as possible. Some crude methods of producing artificial breeze have been used by the farmers, but recently, winnowing-fans, similar to the ceiling fans, have been introduced. These fans may be operated by hand or with a pedal and the number of revolutions are increased by using a pair of gears or with sprocket and chain or a belt. In some cases, e.g. the Nilokheri winnowing-fan, such a drive is not necessary. The size of the fan itself is such that it can produce a good breeze. The object of these fans is to produce a breeze only and, therefore, no sieves are provided.

The winnowers are the machines with a series of three to four sieves to do the preliminary grading. A fan is enclosed in a box-like arrangement and is rotated fast. The sieves are also provided with a shaking arrangement. The steel winnowers cost Rs 1,000 to 1,500 whereas the wooden ones cost Rs 800 to 1,000.

Machines Tools Processing **:** Under this heading, the machines that process the farm produce during the post-harvest period have been described. These machines include sugarcane-crushers, fodder-cutters, groundnut-decorticators, paddy-hullers or paddy-shellers.

Crushers for Sugarcane : The growers of sugarcane cannot dispose of their entire harvest as cane. Part of it has to be converted into *gur* or jaggery. *Gar* is also a popular article of food in the villages. Formerly, the juice was extracted from the cane with crude two-roller wooden crushers. With the growing of improved varieties of cane, a three-roller cast-iron crusher operated either by bullock power or by engine power has been put into the market. It has greater crushing efficiency and, therefore, farmers are either buying such crushers or taking them on rent during the season.

Cutters for Fodder : Of all the improved agricultural implements, the fodder-cutter has become most popular, particularly in Punjab, Haryana, Delhi and Uttar Pradesh areas where green *jouar* or corn is required to be cut for feeding to the cattle. Two sharp steel blades are fixed to a circular flywheel. An automatic pushing arrangement is also provided. In some areas, where power is available, farmers are using the power chaff-cutters or fodder-cutters, or getting their chaff cut on rent basis. Chaff-cutters with good-quality blades are now manufactured in large numbers in India and are available at fairly reasonable prices. The power-driven chaff-cutters can be used also for cutting dry *jowar* stalks, which is not easy to cut with a hand operated chaff-cutter. For farmers, who own small holdings, the *Gopuri* type chaff-cutter designed at Sewagram near Wardha is recommended. It costs only Rs 40 to 50, whereas the cost of the hand-operated circular chaff-cutter is Rs 200 to 300.

Decorticators for Groundnut : When groundnut is to be sown, women labourers are employed for breaking the pods and taking out the kernels. In doing so, the testa of the kernel also gets separated to a considerable extent making it useless for sowing, because such damaged kernels will not germinate. Groundnut-decorticators, therefore, have been designed. The Maganwadi type of groundnut-decorticator is like a cradle with a screen placed below. The other rotary decorticators, either hand-operated or power-operated, are also available. The price of a hand-operated decorticator is Rs 200 to 300 arid the price of a power-operated one is Rs 600 to 1000

Hullers for Paddy : Paddy-hulling or paddy-shelling has been done in India with a *dhenkli* or wooden lever. The Khadi and Village Industries Commission has introduced a Rajasthan type of huller which is quite efficient. Recently, a manufacturer in India has put into market hand-operated hullers similar to Japanese types. They cost Rs 1,000 to 1,500, but are good for small holdings. They are operated by two persons with levered handles. The shelling or hulling is accomplished by the impact of the grain on a very fast-revolving rubber-disc.

A manufacturer has also put on the market a paddy-polishing machine. After the paddy is shelled or hulled, the grain is polished with this machine.

Farm System of Other Nature

Apart from the manual labour, the greatest amount of farm power in India comes from animals. Among the animals, bullocks are used most. The efficiency of using bullocks depends upon their feeding, maintenance, manner of yoking and training. The draught exerted by a work animal is related to its body-weight, the mechanics of working, the method of hitching the implement and the type of harness or yoke used. The following table shows the range of the ratio between the weights of the bullocks and the draught exerted in ploughing.

Body Weights of Bollocks and the Draught Exerted in Ploughing:

Average body-weight per pair of bullocks (kg)	*Locality*	*Draught (kg)*	*Weight draught ratio*
432	Bankura	66.678	6.46
463.592	Tripura	46.449	10.0
453.592	Patna	53.9795	8.33
726.747	Jhansi	59.874	11.95
771.107	Delhi	66.178	11.60
771.107	Aligarh	78.471	9.82
771.107	Coimbatore	146.149	5.20
907.194	Baroda	138.616	6.55
952.553	Karnal	134.535	7.07
1166.670	Indore	183.064	6.00

It may be assumed that the bullocks exert roughly 1/12 to 1/5 of their body-weights as draught. There is reason to believe that the draught animals generally have the ability to exert at least for a short period much greater pull than that indicated in the above table. There are cases on record, where a pair of bullocks weighing 725 to 900 kg has worked mouldboard ploughs having a draught ranging from 145 to 181 kg continually for 30 to 40 days without any adverse effect on their health. That is why sometimes a pair of good bullocks can do a better job than a 4- or 5-hp small tractor, which requires a very considerable amount of its horse power for its own movement. It cannot exert even for a small period appreciably increased draught which a pair of bullocks can exert.

Bullocks, buffaloes, or camels are commonly used in India as work animals. Camels are mostly used in desert areas and also for Persian wheels, oil *ghanis* or *kohlus* and to transport carts in Punjab, Rajasthan and U.P. Since bullocks are far commoner than other animals, some work has been done in determining their horse power. Their power is related to their body-weights. The following table shows the average body-weights of mature bullocks of the main Indian breeds used in farm operations:

Breed	*Average body-weight (kg)*
Nimari	303.907
Nagori	317.515
Siri	317.515
Sahiwal	408.233
Gaolao	408.233
Tharparkar	430.913
Kangayam,	522.601
Deoni	522.601
Haryana	508.023
Krishna Valley	544.310
Kankrei	544.310

Some tests have indicated that the horse power of two bullocks operating a Persian wheel is 0.877 and of those engaged in ploughing varies from 0.29 to 1.87. The walking speed of bullocks working in a circle with a radius of 3 to 4 m, while operating a Persian wheel is 2.5 miles per hour, with a draught of 50 to 51 kg. Bullocks engaged in ploughing walk at a speed of about 1.7 miles per hour and slow down to 1.2 miles per hour, at the end of the day's work. Thus at this speed, a pair of bullocks will plough 0.8 to 0.12 ha in a day with a *desi* plough and 0.2 to 0.4 ha in a day with a medium type of mouldboard plough in an alluvial soil.

Improvement in Yokes and Harnesses: It has long been realised that with improvement in the harness, the yoke and the method of hitching, the implements are likely to lead to a more efficient utilization of draught. Experiments in this connection are going on at Coimbatore, Allahabad, and Bardoli and a number of other places. The experiments conducted at

Allahabad have shown that among the indigenous yokes, the Nagpuri yoke gives the greatest draught. This yoke has now been introduced, at least on an experimental basis, almost all over India and is being manufactured by some of the firms in U.P. The yoke is made of *babool* or *kikar* wood. Because of its shape and curvature, it rests properly on the necks of the bullocks and thus gives a greater draught with comfort to the animal. In case a good yoke is not used, the animal gets a neck gall and thereby reduces its efficiency. The weight of the yokes used in India varies from 7.25 to 15.875 kg depending on the type of wood used and the dimensions compatible with the strength required to withstand the pull exerted by the bullocks. However, most of the two-bullock yokes are good for a pull of 45.359 to 54.431 kg which is the draught commonly required for agricultural implements in use, particularly of the indigenous types. Efforts have also been made to introduce one-bullock yokes, but not with success.

Soil Conservation and Water of Irrigation and Drainage : Soil and water conservation go hand in hand. It is necessary to make the best use of the available rainfall in the scarcity areas, particularly in the districts, like Ahmadnagar, Rayalaseema and Sholapur. Some of the implements which are used for moving the soil or for making bunds have already been described. These implements include the soil scoop, the bund-formers, *keni,* etc. Their description, therefore, is not repeated here. The dry-farming implements, such as a long-bladed *bakhar,* a seed-drill having a distance of 3.81 cm between the tines and specially designed intercultural implements are recommended for dry-farming operations, the principal idea being to utilize, as far as possible, the largest amount of rain water. By bunding the fields and also the *nallahs,* some quantity of water is absorbed and conserved by the soil. This water, in turn, accumulates in open wells or tanks and can then be utilized for irrigation. Another set of implements and machinery used for preparing the land for irrigation consists of a float, a drag, and a leveller. The float consists of a rectangular wooden frame to which cross plates are fixed. The last cross plate is inclined in the direction of the movement of the float. The float is mostly used where the land has already been levelled fairly well and where it is necessary to level it further in order that the irrigation water flows over the land uniformly and does not accumulate in depressions. A drag is an implement which throws the soil in one direction, making the channels flat-bottomed. The leveller is similar to the *keni*, but is longer for the uniform levelling of seedbeds. Water is carried through irrigation channels from wells to the fields and in order to minimize the loss of water through seepage, it is necessary to line them either with bricks or with cement. Recently, the workers of the Indian Road Research Institute have made a cheaper type of chemical out of creosote. This material can be used to plaster earthen water-channels. A kind of weedicide is also mixed with it to prevent weeds from growing in the channels.

One of the major difficulties of the farmers is the raising of water from the wells, tanks or canals to irrigate the fields. Since time immemorial, various indigenous devices have been used for lifting water, the simplest ones being a *dhenkli*, a Persian wheel, a *mhot*, a *pikota*, and *dons*. Where the water level is just near the ground, swing baskets are used. They can lift water from depth of 0.9 to 1.2 m. Two persons are required to work this swing basket. For lifts up to 1.828 m boat-shaped wooden or metal scoops, called *dons*, are used in Orissa, Bihar and Uttar Pradesh. The *dhenkli* consists of a lever and a bucket suitable for lifting the water from wells Up to 4.572 m from the surface of the ground. In the case of a *pikota*, which is common in Tamil Nadu, the operator has to walk on a levered beam. The capacity of the bucket of the *dhenkli* varies from 9 to 18 litres and its water duty, as investigated in Uttar Pradesh is less than a hectare.

The leather *mhot or charsa* is common in Bihar, Uttar Pradesh, Rajasthan, Madhya Pradesh and Maharashtra. The *mhots* in northern India are merely big leather bags, whereas the *mhots* in Andhra Pradesh, Maharashtra and Karnataka have a spout which allows it to be emptied automatically. They are worked by one pair of bullocks and up to depths of 9 to 12 m, but when the lift is more than 15 m, as in Rajasthan, two pairs of bullocks are used. The capacity of the bucket varies from 136 to 181 litres depending on the depth of the well. The discharge is about 6,819 to 9,092 litres per hour and it takes three to four days to irrigate 0.404 hectare. Some manufacturers have also prepared sheet metal *mhots* differing in design and capacity and incorporating into them small improvements like automatic valves in the bottom, self-balancing paired buckets or the bullocks operating in a circle instead of on a slope. The discharge per hour of the improved *mhots* is almost the same as that of the leather *mhots* used by the farmers.

The Persian wheels, or *rakats*, are extremely common in northern India, Uttar Pradesh, Rajasthan, Punjab and northern Maharashtra. Wherever the water level is between 4.5 and 9 m below the ground level, they are popular. A Persian-wheel consists of an endless chain and mild steel flats to which are fixed galvanized-iron buckets. The endless chain rotates on a large cage wheel. The axle of the cage wheel is geared with a pinion to the big gear, which is rotated by a pair of bullocks or a camel walking in a circular track. The modern Persian-wheels are fitted with ball-bearings, and several improved types have been produced. Mention may be made of 'Mayadas' double Persian-wheel which has found some favour in Uttar Pradesh. The average draft of a Persian wheel working up to a depth of about 4 m comes to 59 kg and that working up to a depth of 7.3 m, 86.1 kg. The discharge ranges from 8184.3 to 22730.4 litres per hour and the efficiency ranges from 34 to 44 per cent.

With the advent of engine-power or electric-power, a very large number of Indian farmers are going in for centrifugal pumps. This tendency is invariably growing where there are cash crops, or orchards or gardens. The centrifugal pumps are worked with internal-combustion oil-engines or electric motors. The state departments of agriculture also give *taccavi* loans, and other facilities for the installation of centrifugal pumps. In operating centrifugal pumps, priming is very important, because the centrifugal pump works on the principle of vacuum. The seals of the centrifugal pump should also be tight. In the case of those wells where the water level fluctuates considerably between the rainy season and the summer season and where it is necessary to apply irrigation in summer, additional platforms have to be installed inside the well for fixing the pump at different levels. Because a centrifugal pump has the limitation of raising the water up to the maximum level of 7 m theoretically and up to 5 m practically from the level of the water, the centrifugal pump has to be fixed at a height of 5 m from the water level to get the best out of it.

With the assistance of some of the foreign firms, the Indian manufacturers are also putting into market submersible or floating pumps that rise or go down according to the fluctuating depth of the well. In such cases, it is not necessary to put one or two different plates for fixing the pumps. For very deep wells, the turbine pumps are recommended. These are manufactured specially on order. For tube-wells also, the turbine pumps are useful. In the case of filter-point tube-wells in the coastal region of Andhra Pradesh and Tamil Nadu, the centrifugal pumps can be used, because there the water level is, on an average, 5 m below the ground level. The filter-point tubewells fitted with Such centrifugal pumps are very cheap and, therefore, their use is becoming extremely popular in these regions. The use of a hydraulic ram in the hilly areas is also recommended. It does not require any electricity or mechanical power. The energy that is developed by the falling of the water to a depth 2.438 m is used to raise it to the field level. Such hydraulic rams can be conveniently used in the hilly areas, such as Himachal Pradesh, Manipur, Mizoram and the

Western Ghats. Irrigation water, even if it is readily available, should be judiciously used ; otherwise, it not only damages the crops but also damages the soil to the extent of making it barren. The injudicious use of irrigation water raises the salts in the soil to the surface, making the growth of plants impossible. It is only by the addition of organic matter or of gypsum or both that it is possible to reclaim such lands. Therefore, the extravagant use of irrigation water should be avoided. In those soils where water often accumulates, it is necessary to provide for drainage.

System of Drainage

Good drainage is complementary to irrigation. In those areas where there is plenty of irrigation water, as in Punjab, Haryana and western Uttar Pradesh, drainage has become almost a national problem and is agitating the minds of the farmers and of the government. It is, therefore, necessary to know the methods developed for draining away the excess water from the fields. Water accumulates in the soil either from rains or through seepage from canals or irrigation channels and reservoirs. The high subsoil water-table damages agricultural land in many ways, leading particularly to the accumulation of undesirable salts in the root-zone or on the surface of the soil, involving drainage problems. The root cause of the drainage must be thoroughly investigated and understood. The problem of persistent water-logging is often complex and expert advice might be sought to solve it. In areas of heavy rainfall, good surface drainage has to be arranged. The drainage ditches should be wide and of trapezoidal or parabolic shape, so that they offer a rapidly increasing cross-section to the flow of the stream as the water level in them rises. On relatively flat land, they may be made to run down the slope. The velocity of water in the drains should be maintained at non-erosive level by proper grading and shaping. The spacing of the field drains should be worked out taking into consideration the character of the soil, the slope, the intensity of the rainfall, etc. A field drain should be laid to join a utility channel of adequate dimensions. These utility channels should be sufficiently big to take the maximum quantity of water.

In those soils where water accumulates underground within a depth of 0.6 to 1.2 m, underground drainage is needed. In case this high water-table is due to percolation from an external source, intersection drains should be installed to catch and divert the water. There are several types of underground drainage systems, but most of them are costly. As the work is of a permanent nature, it involves considerable expense. A master plan for draining the land should be prepared in consultation with specialist engineer and after making a preliminary survey of the levels. The best kind of drainage pipes are hard, burnt, earthenware flange-less pipes. The diameter of the pipe is usually 10 cm. The pipes should be laid well below the lowest root level of the crops proposed to be grown on the land. The joints of the pipes are important and should be left uncemented, but protected with a strip of tarred paper to prevent the entry of silt or soil, as the drainage of water takes place largely through the joints and not through the porous body as is usually thought. The spacing of the drains depends on the level of the land and the permeability of the soil. The drains can be placed in a grid or herring-bone fashion, depending upon the level of the water-logged area. In the case of sugarcane-growing areas in Andhra Pradesh and Tamil Nadu, the farmers are using mole drains. They are made at a depth of 0.45 to 0.6 m and are quite effective. The tractor-mounted trenching-machines are very useful in digging the channels for putting the tile drains.

In order to drain the water from the fields, the Japanese farmers use vertical pumps having low lifts, and high volume. Samples of such pumps were recently obtained from Japan and suitable designs were prepared. Such pumps are now available in India. The capacity of such pumps is 136,380 to 181,540 litres per hour and they can be effectively used for draining the soil, if the utility channels are available.

Bibliography

Anderson, J.R. : *A Handbook of Agriculture,* W.M.C Brown Co., Iowa, 1970,

Bhatia, B.M. : *Poverty, Agriculture and Economic Growth,* Vikas Publishing House, New Delhi, 1977.

Bose, N.K. : *Rural Life in India,* National Book Trust, New Delhi, 1971.

Boserup, E. : *The Conditions of Agricultural Crops,* Oxford, London, 1965.

Buchanan, K. : *India: The Land and People,* Crown, New York, 1981.

Champion, H.G. : *Indian Forest Records,* Vikas Pub., New Delhi, 1936.

Colinvaux, Paul A. : *Introduction to Agriculture,* John Wiley and Sons, New York, 1973.

Cressey, G.B. : *Asia's Land and Peoples,* McGraw-Hill, New York, 1963.

Dey, A. K. : *India's Mineral Wealth,* Oxford University Press, New York, 1990.

Dharma, M. : *Fisheries Development in India : Some Aspects of Policy Management,* Concept Publishing Co., New Delhi, 1985.

Dhillon, S.S. : *Agricultural Geography,* Tata McGraw-Hill Pub. Co., Ltd. New Delhi, 1984.

Edgar, S. : *Location of Agricultural Production,* University of Florida Press, 1954.

Ehrilich, P.R. : *Eco-Science,* W.H. Freeman & Co., San Franciscess, 1977.

Enyedi, Gyorgy : *Geographical Types of Agriculture,* Applied Geography in Hungary, Budapest, 1964.

Gandhi, M.P. : *Major Industries of India, 1970-71,* M.P. Gandhi and Co., Bombay, 1972.

Ganguli, B.N. : *Trends of Agriculture and Population in the Ganges Valley: A Study in Agricultural Economics,* Methuen, London, 1938.

Gerald Brian P. : *Agriculture in India,* Oxford University Press, London, 1974.

Gregory, D. : *Ideology, Science and Human Geography,* Hutchinson, London, 1978.

———: *The Geographical Imagination, Social Theory and Human Geography,* Hutchinson, London, 1985.

Hartshorn, T.A. : *Interpreting the City: An Urban Geography,* John Wiley & Sons, New York, 1990.

Hathaway, D.E. : *Government and Agriculture: Public Policy in a Democratic Society,* MacMillan, London, 1963.

Hourani, G.F. : *Arab Seafaring in the Indian Ocean and Early Medieval Times,* Princeton University Press, Princeton, 1951.

Howard, F. : *Geography of Agriculture: Themes in Research,* Prentice-Hall, New Jersey, 1970.

Jenkins, M. M. : *The Himalayan Kingdoms: Bhutan, Sikkim and Nepal,* Van Mostrand Company, New Jersey, 1963.

Joshi, H.L. : *Agricultur,* Rawat Publications, Jaipur, 1990.

Joshi, J.K. : *Social Structure and Cultural Change in a Bhil Village,* Heights, New Delhi, 1974.

Kumar, V.K. : *Kanpur City: A Study in Environmental Pollution,* Tara Book Agency, Varanasi, 1982.

Lyman, H. : *Beginnings of Agriculture in India,* Vikas Pub., New Delhi, 1968.

Major, John, S. : *The Land and People of India,* Lippincott, Philadelphia, 1989.

Majumdar, D.N. : *Races and Cultures of India*, Asia Publishing House, Bombay, 1958.

Mansfield, G.B. : *Farming Systems of the World*, Praeger, New York, 1970.

Miracle, M.P. : *Traditional Agricultural Methods in India*, Food Research Institute, New York, 1964.

Morgan, W.B. and R.J. Munton : *Agricultural Geography*, Methuen & Co. London, 1972.

Mukherji, Shekhar : *Land-Use Change, Environment and Sustainable Development*, Banaras Hindu University, Varanasi, 2000.

Nicholson and Max : *The Environmental Revolution*, McGraw-Hill, New York, 1970.

Oliver, D. L. *Native Cultures of the Pacific Islands*, University of Hawai Press, Honolulu, 1989.

Parain, C. : *The Evolution of Agricultural Technique* Cambridge University Press, Cambridge, 1966.

Parmer, B.H. : *The Green Revolution in India*, The MacMillan, London, 1980.

Pedelaborde, P. : *The Monsoon*, Methuen and Co., London, 1963.

Persson, G.A. : *Resources and World Development*, John Wiley, Chichester, 1987.

Pritchard, J.M. : *Landform and Landscape in India*, Edward Arnold, London, 1979.

Pyke, M : *Man and Food*, McGraw Hill, New York, 1970.

Qureshi, M.H. : *India : Resources and Regional Development*, N. C. E. R. T., New Delhi, 1990.

Raghava, S. P.S. and Dadiani, N.K. : *The Hindu Survey of Indian Agriculture*, Kasturi & Sons Ltd., Chennai, 1999.

Raychaudhuri, S.P. : *Land and Soil*, National Book Trust of India, New Delhi, 1966.

Rotberg, R. and Rabb, T. : *Hunger and History: The Impact of Changing Food Production and Consumption Patterns on Society*, Cambridge University Press, New York, 1984.

Sengupta, S. : *Geography of India*, Concept Pub. Co., New Delhi, 1992.

Shafi, M. : *Agricultural Productivity and Regional Imbalances—A Study of Uttar Pradesh*, Concept Pub. Co. : New Delhi, 1983.

Shaft, M. : *Ganga-Yamuna Doab, in India : Regional Studies*, National Committee for Geography, Calcutta, 1923.

Singh, G.B. : *Transformation of Agriculture*, Vishal Publication, Kurukshetra, 1979.

Singh, J. : *An Agricultural Atlas of India: A Geographical Analysis* Vishal Publication, Kurukshetra, 1974.

Singh, Jasbir : *An Agricultural Atlas of India : A Geographical Analysis*, Vishal Publications, Kurukshetra, 1974.

Singh, R.L. : *India: A Regional Geography*, National Geographical Society of India, Varanasi, 1971.

Singh, Savindra and Upadhyay, D.P. : *Facets of Agriculture*, Thinkers Library, Allahabad, 1985.

Singh, Surendra : *Agricultural Development in India : A Regional Analysis*, Kaushal Publications, Shillong, 1994.

Small, R. J. : *The Study of Landforms*, Cambridge University Press, London, 1970.

Sundaram, K.V. : *Planning Revisited: Rural Area Development in India: Vision and Reality*, R. N. Dubey Memorial Lectures, Allahabad, 1999.

Sundaram, K.V. : *Regional Planning and National Development*, Vikas Pub. House Pvt. Ltd., New Delhi, 1978.

Taylor, James A. : *Weather and Agriculture*, Pergamon, Oxford, 1967.

Thomas, W.L. : *Man's Role in Changing the Face of the Earth*, University of Chicago Press, 1965.

Wadia, D.N. : *Minerals of India*, National Book Trust, New Delhi, 1994

Wadia, F.K. : *Agro-Climatic Regional Planning in India*, Concept Publishing Company, New Delhi, 1996.

Zelinsky, W. : *A Textbook of Agriculture*, Prentice Hall, New Jersey, 1966.

Index

A

Agricultural, 21.
Agricultural Activity, 135.
Agricultural Agencies, 102.
Agricultural and Horticultural, 191.
Agricultural and Horticultural Crops, 191.
Agricultural Aviation, 253.
Agricultural Commodities, 24.
Agricultural Departments, 156.
Agricultural Development, 18, 24, 78, 104., 156.
Agricultural Development Programme, 24, 156.
Agricultural Electronics, 253.
Agricultural Engineering, 249, 250, 252.
Agricultural Engineering Colleges, 251.
Agricultural Engineering Section, 252.
Agricultural Engineers, 254.
Agricultural Extension Officer, 252.
Agricultural, Forestry, 23.
Agricultural Implements, 250, 251, 254, 271, 275.
Agricultural Land, 23, 279.
Agricultural Limestone, 218.
Agricultural Machinery, 253.
Agricultural Machines, 249.
Agricultural Meteorolog, 156.
Agricultural Meteorology, 151, 156.
Agricultural Officer, 252.
Agricultural Operations, 156, 254.
Agricultural or Forestry Products, 24.
Agricultural Planning, 6.
Agricultural Potential, 137, 151.
Agricultural Practices, 24.
Agricultural Produce, 22, 126.
Agricultural Production, 16, 45, 111, 112, 249, 252.

Agricultural Productivity, 151, 250.
Agricultural Purpose, 78, 158.
Agricultural Regions, 112.
Agricultural Research, 113, 254, 256.
Agricultural Science, 126.
Agricultural Scientists, 126, 156.
Agricultural Season, 262.
Agricultural Statistics, 5, 6, 9, 20.
Agricultural Universities, 113, 251.
Agricultural Uses, 2, 12, 15, 20, 21.
Agricultural Wealth, 82.
Agricultural Workshop, 263.
Agricultural Year, 18, 22.
Agricultural Yields, 147.
Agriculture, 24, 31, 77, 79, 80, 97, 110, 111, 136, 137, 138, 150, 195, 237, 239, 249, 250, 251, 253, 256, 278.
Andaman and Nicobar Islands, 20.
Andhra Pradesh, 3, 4, 5, 11, 20, 21, 22, 23, 86, 88, 90, 99, 115, 116, 117, 118, 119, 120, 121, 123, 124, 125, 140, 146, 150, 257, 262, 266, 277, 278, 280.
Arabian Sea, 29, 30, 136, 140, 143.
Arunachal Pradesh, 4, 5, 20, 21, 115, 138.
Assam, 3, 4, 5, 10, 20, 21, 23, 30, 31, 33, 69, 83, 90, 111, 112, 115, 123, 124, 138, 144, 146, 152.
Australia, 12, 14, 249.

B

Bay of Bengal, 29, 30, 31, 136, 137, 142, 143.
Bihar, 3, 5, 11, 20, 21, 22, 23, 83, 88, 89, 92, 99, 111, 115, 117, 121, 122, 123, 124, 125, 139, 142, 143, 145, 160, 161, 162, 258, 259, 276.
Bombay, 8.
Brazil, 14.
Burma, 12, 13, 14, 265.

C

Calcutta, 144.
Canada, 12, 14, 154, 249.
Cation Exchange Capacity, 175.
Chandigarh, 139, 151.
Chemical Methods Reclamation, 181.
Chennai, 155.
Chhotanagpur, 88, 89, 125.
China, 4, 5, 14, 265.

Classification of Essential Nutrients, 188.
Classifying Soil Water in Biological Terms, 173.
Coimbatore, 258, 262, 273, 274.
Components of Soil, 165.
Compost, 182.
Cropping Pattern, 102, 104, 112, 114, 119.
Cultivating Areas, 22.

D

Defining Soil, 78, 106, 157.
Delhi, 4, 20, 22, 99, 121, 139, 151, 155, 253, 260, 271, 273.
Density of Soil, 168.

E

Egypt, 14.
Europe, 15, 111.

F

Farm System, 272.
Field Capacity, 174.
France, 12, 14.

G

Germany, 14.
Glossary, 190.
Goa, 4, 20, 21, 88.
Green Manuring, 182.
Green Manuring : Merits, 183.
Gujarat, 3, 4, 5, 11, 20, 21, 22, 23, 30, 31, 85, 86, 98, 116, 117, 118, 119, 120, 121, 123, 138, 139, 150, 152, 161.

H

Haryana, 3, 4, 20, 22, 23, 85, 92, 98, 113, 116, 120, 121, 122, 123, 125, 139, 151, 152, 162, 263, 268, 269, 271, 274, 279.
Himachal Pradesh, 3, 4, 5, 20, 21, 23, 112, 116, 117, 118, 139, 143, 151, 176, 177, 278.
Hygroscopic Coefficient, 175.

I

Impact on Soil Temperature, 171.
India, 1, 4, 6, 9, 10, 11, 12, 13, 14, 15, 17, 18, 19, 20, 21, 22, 23, 25, 29, 30, 31, 33, 35, 37, 48, 58, 72, 77, 81, 82, 88, 90, 94, 97, 98, 103, 109, 111, 112, 113, 114, 115, 116, 117, 118, 119, 120, 121, 122, 123, 124, 126, 136, 137, 138, 142, 143, 144, 145, 146, 147, 148, 149,

151, 152, 153, 154, 155, 156, 160, 182, 249, 250, 251, 253, 254, 255, 256, 259, 260, 261, 265, 266, 267, 268, 270, 271, 272, 273, 275, 277, 280, 296.
Indian Agriculture, 17, 150, 251, 253.
Indian Bladed, 259.
Indian Breeds, 274.
Indian Civilization, 39.
Indian Coasts, 143.
Indian Council, 254, 256.
Indian Council of Agricultural, 113.
Indian Crop-production, 111.
Indian Farmer, 154, 250, 254, 255, 256, 267, 268, 269, 277.
Indian Forests, 15.
Indian Industries, 250.
Indian Manufacturers, 278.
Indian Meteorological Department, 144, 145, 150, 155.
Indian Ocean, 30, 136.
Indian Peninsula, 30, 112.
Indian Road Research Institute, 276.
Indian Soils, 68, 69, 72, 73.
Indian Subcontinent, 29, 30.
Indore, 258, 266, 273.
Irrigating Areas, 18.
Italy, 14, 265.

J

Jammu and Kashmir, 5, 11, 20, 23, 116, 117, 137, 139, 151.
Japan, 12, 13, 14, 249, 253, 265, 280.

K

Karnataka, 3, 4, 5, 20, 21, 22, 23, 87, 88, 89, 90, 91, 98, 112, 115, 116, 117, 118, 119, 120, 121, 123, 124, 125, 140, 150, 152, 262, 266, 277.
Kerala, 3, 4, 5, 11, 20, 22, 23, 30, 85, 90, 91, 99, 112, 115, 116, 124, 137, 140, 160, 161, 258.
Kolkata, 155.

M

Madhya Pradesh, 3, 4, '10, 11, 20, 21, 22, 23, 85, 86, 87, 88, 89, 90, 98, 111, 112, 115, 116, 117, 118, 119, 120, 121, 122, 124, 125, 139, 143, 148, 150, 152, 276.
Madras, 10, 11, 144, 160, 161, 263.
Maharashtra, 3, 4, 20, 21, 22, 23, 69, 86, 87, 88, 90, 91, 98, 116,

118, 119, 120, 121, 122, 123, 124, 125, 139, 148, 150, 160, 161, 257, 259, 261, 262, 266, 276, 277.
Major Elements, 180.
Malaysia, 265.
Manipur, 3, 4, 20, 21, 23, 115, 138, 278.
Meghalaya, 3, 4, 20, 21, 23, 115, 138.
Mizoram, 4, 20, 21, 115, 138, 278.
Mumbai, 155, 258.
Mysore, 11, 89, 91, 150, 151, 160.

N

Negative Impact of Acid Soil on Plant Growth, 178.
Night Soil, 182.
Nitrogen Loss in the Field, 186.

O

Organic Manures, 181.
Orissa, 4, 5, 11, 20, 21, 22, 23, 84, 88, 90, 92, 98, 115, 116, 119, 123, 124, 139, 160, 257, 258, 276.

P

Pakistan, 4, 5, 12, 13, 14, 144.
Pattern of Usage, 2.
Phycobilin, 247.
Poland, 14.
Poona, 151, 156.
Porosity of Soil, 169.
Punjab, 4, 8, 10, 11, 20, 22, 23, 30, 69, 85, 92, 98, 111, 113, 116, 117, 118, 119, 120, 121, 122, 123, 124, 125, 139, 148, 151, 152, 161, 259, 261, 268, 269, 273, 277, 279.

R

Raipur, 85, 89.
Rajasthan, 4, 5, 21, 22, 23, 30, 31, 33, 37, 69, 92, 98, 111, 112, 116, 117, 118, 119, 120, 121, 122, 124, 136, 138, 139, 144, 147, 151, 152, 161, 272, 273, 276, 277.
Relevance and Preparation of Vermicompost, 183.

S

Saline and Alkaline, 161, 179, 183.
Sample Information Data, 105.
Soil Categorization, 81.

Soil Consistence, 170.
Soil Fertility, 32, 78, 188.
Soil Formation, 48, 162, 164.
Soil Formation, 49, 51, 52, 53, 162, 219, 240.
Soil Heat, 171.
Soil Management, 63, 64, 102, 103, 104, 147, 195.
Soil Profile, 52, 53, 54, 83, 102, 158, 163, 164, 221, 233, 235.
Soil Structure, 167.
Soil Texture, 166.
Soil Water Concept, 172.
Soil's Colour, 170.
Soils in India, 98, 160.
Soils of India, 160.
Soil's Temperature, 171.
System of Drainage, 279.

T

Tamil Nadu, 4, 20, 21, 22, 23, 33, 69, 84, 86, 87, 88, 90, 91, 99, 112, 115, 116, 117, 119, 120, 123, 124, 125, 137, 140, 146, 147, 148, 276, 278, 280.
Tripura, 4, 20, 21, 23, 115, 124, 138, 273.

U

U.S.A., 14.
U.S.S.R., 14.
UK, 12, 13, 249.
USA, 12, 13, 154, 155, 249, 259, 265.
Use of Soil, 129, 261.
USSR, 12, 13, 249.
Uttar Pradesh, 4, 10, 11, 20, 21, 22, 23, 69, 84, 89, 98, 112, 115, 116, 117, 119, 120, 121, 122, 123, 124, 125, 143, 151, 153, 259, 261, 262, 264, 269, 271, 276, 277, 279.
Uttaranchal, 139.

V

Various Classes, 25, 62.

W

West Bengal, 4, 5, 7, 11, 20, 21, 22, 23, 30, 31, 83, 88, 89, 90, 91, 98, 113, 115, 121, 123, 124, 138, 139, 146.
Wilting Coefficient, 174.
World, 12, 13, 29, 81, 111, 137, 150.

□□□